# Maneuverability And Safety Of Ships

CARMINE GIUSEPPE BIANCARDI, PhD, CEng

Professor of Maneuverability and Safety of Ships

Library of Congress Control Number: 2012914257
ISBN-10: 1478302429
ISBN-13: 978-1478302421

# DEDICATION

I would like to express my appreciation towards my family, my friends and my colleagues for their supportive encouragements throughout the preparation of this book. Moreover, I would dedicate this book to the very many seamen, captains, engineers, academicians and experts that over many years and with their wisdom and passion have learned me a great deal on the ship controllability

# CONTENTS

Appendix 4: Comparison between calculation and model-scale measurements of the hydrodynamic sway force Y and Yaw moment N evaluated by various facilities

Appendix 5: Comparison od ship simulation and free running model testing of a Jackclass type model with a double vectwin rudder system

Appendix 6: The ship's life cycle

Appendix 7: PREVENT-IT safety methodology for maneuverability

Appendix 8: Design changes of case example 1

Appendix 9: Calculation of risk for case example 1

Appendix 10: Definition of symbols

Appendix 11: Main geometric parameters of the ships stored in the database

# ACKNOWLEDGMENTS

I would like to express my sincere thanks to Professor Chengi Kuo, Department of Ship and Marine Technology, University of Strathclyde, Glasgow, U.K., for his interest, discussions and constructive criticism. It has been a pleasure to have scientific discussions with an individual who is continually open to new ideas and whose academic research is directed toward relevant applications as well as scholarship.

It is impossible to acknowledge all those persons who over the time of this research have helped directly or indirectly. However, a special note of thank is entitled to Professor David R. Dellwo of the U.S. Merchant Marine Academy, Kings Point, New York, U.S.A, who influenced to shape my thinking towards the classification of maneuvering characteristics.

Moreover, I would also like to acknowledge the followings:

• The members of Panel H-10 (Ship Controllability} of SNAME (The Society of Naval Architects and Marine Engineers) for their contribution of useful information;

• The INSEAN (The Towing Tanks of Rome), Italy for supporting part of the research needed for this book.

• The technical staff of the Department of Ship and Marine Technology, University of Strathclyde, Glasgow, Scotland, UK for their competent help in running the experimental trials;

• Captain Alexander van Baalen for providing the Jackclass Type model;

• The Lecturers and the staff of the Department of Applied Sciences of the University of Naples "Parthenope", Italy for their friendly cooperation and support during the years.

• Ms. Giulia Costa (http://ilcuratoreeditoriale.wordpress.com/) for her support in editing this book

# ABSTRACT

The book begins by performing a critical review of existing approaches for dealing with Ship Maneuverability and Ship Safety before considering fresh understandings of these terms. Different attempts to integrate Safety with Maneuverability are then examined. The weak features of these attempts are considered and scope for developing fresh approaches is then presented. This is followed by an explanation of alternative new ways of treating a ship's maneuverability. In order to meet the requirements for operational specifications and safety in a cost-effective way, the book proposes an approach for relating Safety with Maneuverability in the appropriate phases of a ship's life cycle. The approach is based on a preventive Safety Methodology while introducing fresh indices and criteria for assessing the ship's maneuverability. Two case examples are used to show how the methodology can offer an effective approach for designing and operating ships that can meet improved maneuvering-safety requirements. One case example studies a dry cargo ship of a Mariner Type while the other studies a service ship of a Jackclass Type. Free running model testing is used to verify the results of the Jackclass type.

# 1 INTRODUCTION

One safety aspect of ships is aimed in preventing contacts and collisions during maneuvering operations.

Statistics show that collisions and contacts are one of the most common hazards of seagoing ships, [1-5] and poor maneuverability has been called as a significant element in some major accidents, [5, 6]. Moreover, a ship's own maneuverability may limit the range of operability. So, poor maneuverability can impair both safety and efficiency of ships. Furthermore, the maritime industry customers will always be specifying higher maneuvering qualities, coupled with cost effectiveness and an acceptable level of safety in relation to both human life and ocean environment. Putting high pressure on technically trained staff to innovate, the maritime industry management is seeking technical solutions to these requirements. However, management is interested in profits and the return on investments, [7]. If customers can be convinced that the implementation of improved maneuverability and safety is cost-effective, i.e., that they will reduce operating expenses and increase profit, it is more likely they would support the maneuvering-safety program as a solution to their requirements.

To achieve maximum commercial benefits such as payload and fuel economy, ship geometric parameters are fixed at the preliminary design stage. Designs obtained in such a way may not possess desirable maneuvering and safety characteristics. Even tough, a ship designer is willing to consider maneuverability at the design stage; some frustrations can limit his aspirations. At the design stage, model scale testing may either or not included. Assuming that the model testing is contemplated, prior the experiments there could be two typical circumstances. One is at a point of the design stage where many hull parameters remain subject to change. The other is the case where model testing is not intended until later stage of the ship design or during construction, where all or most parameters are already established. However, there are some cases in which the estimation of the maneuvering performances is required without model testing. In any case, despite the ability to theoretically or experimentally predict maneuverability of a vessel during design has increased over the years, [8-16], some problems on the reliability of the methods for assessing and implementing required maneuverability still remain unsolved, [16-18]. This situation would seem to place maneuvering qualities in a secondary position in the ship's life cycle. Sometimes, certain maneuvering appendages are added to the hull to compensate poor maneuverability or to achieve specific operating conditions.

On the other hand, maneuverability is a significant part of ship's safety, [1, 3, 17, 19, and 20]. Indeed, there are international efforts of IMO (international maritime organization) with the objective to improve the ship's Maneuvering-safety, [19]. However, safety is a broad idea, the importance of which is appreciated by everyone, but understandings of its actual meaning tend to vary considerably:

• practicing engineers believe safety is concerned with design, rules and regulations.

• academics think "reliability studies" and "risk analysis" are most relevant to safety.

• operators would regard operating procedures as most closely associated with safety.

None of these approaches alone seem consistent with enhancing maneuvering-safety features of ships. On the other hand, the problem of deciding how to discriminate between ships with acceptable and unacceptable maneuvering-safety features still remains unsolved. To provide a proper solution, safety should be incorporated with maneuverability over the ship's entire life cycle.

Success in incorporating safety with maneuverability can generate at least the following benefits:

• improve operational performances;
• reduce cost of training;
• reduce down-time for repairs or modifications;
• decrease losses through failure to complete contracts on time;
• reduce economic loss to employee's family.

In other words, it means to provide an approach that satisfies with equal effect the three separate sets of criteria:

• Operational specifications;
• Safety;
• Cost Effectiveness.

Generally, using the most advanced technology without due regard to cost; it is relatively straightforward to meet a demanding maneuvering specification. Or alternatively, for the operation to have a very high level of safety if the efficiency of the operation is not a concern. The real challenge, here, is to meet all three sets of criteria simultaneously. In the past, there has been a vague idea of what relating safety with maneuverability means. Looking to specific aspects of the problem – say operational procedures or constraining restricted area to ship with specific maneuverability – some studies, [12, 18, 21-24], have suggested attempts to associate safety with maneuverability. Indeed, most of the available approaches are a scatter collection of assessing, measuring or predicting maneuvering performance of ships at different phases of its life cycle. Unfortunately, none of the available approaches relate safety with maneuverability during the entire life-cycle. There are some tentatives of guiding the maneuvering assessment process at the design stage or to provide information to the personnel. Lately, IMO (international maritime organization), [25], has been seriously elaborating full-scale maneuvering standards. Because of these standards, a solution is definitely required for ensuring. That the compulsory level of maneuvering-safety is achieved over a ship's entire life-cycle.

**A number of possible solutions could be explored**.

However, the relationships between the variables cannot be defined by analytical equations or numerical relations because they are closely inter-related and highly complex. Moreover, some of the elements involved are qualitative rather than quantitative. The way forward is based on a fresh approach.

After devising a common understanding of the terms "maneuverability", "safety" and "what are acceptable maneuvering qualities", in writing this book the efforts have been directed to devise an approach that relates safety with maneuverability while satisfying the three sets of criteria. Such an approach is based on the understanding that safety is linked with maneuverability over the appropriate phases of a ship's life-cycle. The approach combines alternative maneuvering criteria and a safety methodology within the same framework. The approach is based on the PREVENT-IT Safety Methodology, [26], while introducing sensitivity indexes for assessing the ship's maneuverability. Examples are provided to illustrate the practical applications.

# 2 AIMS

The aims of the book are:

(a) To make a critical review of the available approaches used in treating ship maneuvering and safety.

(b) To examine possible ways of relating safety features with ship maneuverability.

(c) To put forward an approach that incorporates safety features with ship maneuverability and to illustrate its application with practical examples.

(d) To give consideration to the application of the proposed method in practical situations and requirements for its further developments.

# 3 APPROACHES TO MANOEUVRABILITY

The term "Maneuverability" can have different interpretations. Several sources, [13, 19, 20, 23, 24, 27-30], provide sparse definitions of Maneuverability. One of the most accepted understanding is supplied by [8]:

"The controlled change in the direction of motion (turning or course changing). Interest centers on the ease with which change can be accomplished and the radius and distance required to accomplish the change".

However, in the present context, it would be helpful to begin by contributing with a fresh understanding of the word "Maneuverability":

"The quality which determines to what extent a marine vehicle meets the requirements for changing or maintaining a specific direction and speed in a set of environmental conditions"

There are a number of words in that statement calling for some explanations:

Quality: Maneuverability is not something absolute, but a QUALITY to be specified according to given a ship characteristics and operational circumstances. For example, a stopping distance of 7 ship lengths for one ship can be regarded as good, but may not be acceptable for another ship.

Requirements: different operational conditions can require different level of maneuverability.
Maneuverability has to satisfy the demand of operational conditions as well as those of regulatory bodies and human operators.

Environmental conditions: Maneuvering performances are affected by environmental conditions such as the intensities of wind, current, wave, and water depths.

The study of the maneuverability is generally made in two directions. One is to assess maneuverability with a physical model either at full or reduced scale. The other is to assess maneuverability with a theoretical model, i.e. a set of equations of motion, in which the equations' coefficients can be calculated either by means of a theory or by means of a physical model. Both methods are useful to ship designers and ship operators.

The objective of this chapter is to present the different ways of describing and assessing the maneuverability of ships. There are five parts. Each part describes one of the five main approaches to maneuverability, as follows:

- Mathematical Model Techniques and Theoretical Prediction
- Maneuvering Simulators;
- Full Scale Trials;
- Model Experiments;
- Statistical Analysis.

## 3.1. **Mathematical Model Techniques and Theoretical Prediction**

**Goal**:
To describe and assess the maneuverability by means of analytical methods.

**Description**:
The basic dynamics of maneuvering can be described and analyzed using Newton's equations of motion in six degrees of freedom. However, to simplify the study of maneuverability, restriction to the horizontal plane is often used with accurate results, [15]. Basic equations in the horizontal plane can be considered first with reference to one set of axes fixed relative to the earth and a second set fixed relative to the ship. In the equations of motion, the hydrodynamic forces and moments are normally described by a Taylor expansion (see Appendix 2) about the equilibrium condition, resulting in different terms (generally called hydrodynamic coefficients or derivatives) necessary to describe the ship dynamics. One of the most classical approaches is described in [31, 32]. In spite of the apparent simplicity of the equations, the motion of a ship is more conveniently expressed when referred to the axes fixed with respect to the moving ships, see Figure 1. For completeness, the differential equations of motion in the horizontal plane are:

$$(m + m_x)\dot{u} - (m + m_y)vr = X(u, v, r) \qquad \text{(Surge)} \qquad (3.1a)$$

$$(m + m_y)\dot{v} - Y_{\dot{r}}\dot{r} + (m + m_x)ur = Y(u, v, r) \qquad \text{(Sway)} \qquad (3.1b)$$

$$(I_z + J_z)\dot{r} - N_{\dot{v}}\dot{v} = N(u, v, r) \qquad \text{(Yaw)} \qquad (3.1c)$$

The terms in the equations are summarized in Figure 1 and in Appendix 10.

Once the set of the differential equations is derived, the hydrodynamic coefficients have to be determined.

Theoretical methods exist for calculating the hydrodynamic coefficients, [8, 36], however their application is constrained by the main assumptions that:

• the fluid is assumed irrotational;

whilst the following effects can be considered negligible:

• lifting effects,
• free-surface effects,
• viscous stresses and separation,
• propeller-hull-rudder interaction.

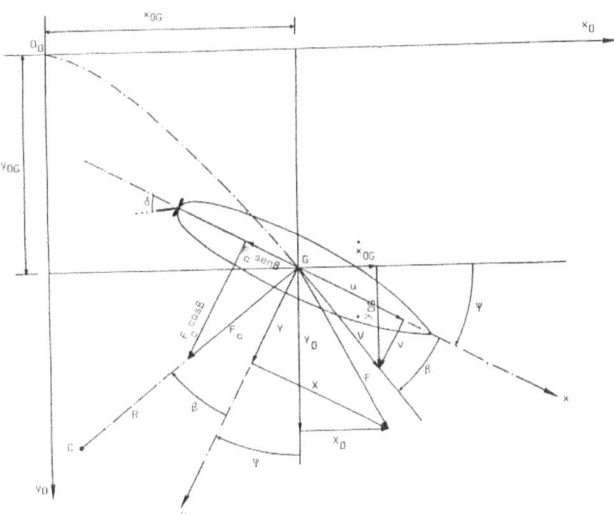

Fig. 1
Body-fixed and space-fixed coordinate systems. The body fixed
system is located at the ship's center of gravity G

On the other hand, other approaches have been developed to take into consideration the presence of the free surface effects. Typical approaches include the two dimensional and three-dimensional methods. Two dimensional methods have been successful applied, [33-34], with significant results. A three dimensional method has been presented by Mikelis, [16, 35]. However the method requires further investigation, [35].

On the other hand, the coefficients can be successfully calculated with a physical model using the technique described in 3.4. (Model Experiments).

**Advantages**:
When the coefficients have been calculated [36], the method permits different maneuvering performances to be readily assessed by calculation. It is possible to introduce different speeds and rudder angles as well as various set of environmental conditions. The method provides means of evaluating maneuvering qualities or operational characteristics of proposed design well in advance of construction. Moreover, it contributes simple means of studying effects of practical variations of basic design which may be necessary to improve performance. The method can be used to determine effects of various environmental conditions on performance. It can be used to study overall performance of various combined systems.

**Drawbacks**:
The equations of motion require an accurate set of ship hydrodynamic coefficients in order to provide reliable maneuvering assessment. The accurate calculation of the hydrodynamic coefficients requires expensive facilities, see section (3.4.). However, theoretical calculation of the coefficients may be more cost-effective. On the other hand, the accuracy is constrained by the assumptions of the coefficients' theory, particularly for maneuvers with large rudder angles. Indeed, in [8], it is shown that theoretical assumptions lead to a reasonable correlation with experimental data for small motions within the linear range. Moreover, it has been demonstrated that theory seriously over-predicts some maneuvering effects, [8].

**Comments**:
Mathematical model technique has often been used since the wide availability of high speed computers. This technique provides a solution and basic understanding of the dynamics of submerged and floating bodies. However, the differential equations of motion are comprised to a number of coefficients or derivatives which are of hydrodynamic origin. Consequently to obtain solutions for any given configuration, it is necessary to know the

coefficients with a reasonable accuracy. Several attempts have been made in the past to fulfill this requirement by utilizing various experimental techniques, theoretical calculation or a combination of both of them, [8, 9, 11, 16, 22, 33, 34, 36, 37-39, 40-42]. However, there are very few theoretical methods available to compute the hydrodynamic coefficients, especially terms involving viscous flow.

On the other hand, standard procedures, used by most ship maneuvering testing facilities capable of conducting captive model experiments (physical model on scale), involve measuring the hydrodynamic forces and, subsequently predicting ship maneuvering characteristics using the equations of motion.

### 3.2. Maneuvering Simulators

**Goal**:
To permit the study of ship maneuverability with allowances, reruns and in some cases with the direct inclusion of human responses by means of models representing the ship-operator-environment system.

**Description**:
A ship in navigation is a man-machine system that interacts with the environment. Maneuvering simulation provides a representation of the maneuvering ship using a mathematical model or a compound of models or a physical model representing that system and environment, [11, 42-44]. Maneuvering simulators are generally divided in two main types:

- hydraulic simulators; and
- computer simulators.

The first type makes use of physical scale models.

Harbor and vessel waterways system accurately modeling the hydraulic flow can be constructed and then free running ship models can be piloted either remotely or with human operators on-board.

A computer hardware and its relative software developed for the scope of representing ship maneuvering behavior is often regarded as a computerized ship maneuvering simulator. There are various types of such a simulator that may greatly differ in accuracy and complexity. Each type has a specific structure of the aforementioned model. Ship maneuvering simulators may generally be divided in four group types:

- Full mission, real-time simulator;
this type is probably one of the most complete ship maneuvering simulator. It includes a replica of a typical ship bridge including outside scene view. Most of the maneuvering situations for research and training can be simulated in real-time condition in a real-looking environment.

- Mini and micro simulator;
it doesn't include the bridge replica, and it is in general a simpler version of the previous type. It is less suitable for some types of research and training.

- Fast-time simulator;
it is a compound of isolated models and procedures. Often, an auto-pilot represents the human controller, for instance. It is usually regarded as an adequate tool for feasibility studies. It may well be very useful to ship designers.

- Part task simulator;
it is a simulator representing only a model of a ship's subsystem. It is a tool for training a specific task or aspect of handling a ship. For example, a radar simulator used for training officers to collision avoidance.

These types of maneuvering simulators are controlled by computers employing mathematical models of varying complexity to predict the ship's trajectories. Effects of shallow water, banks, currents, wind, waves, and tugs are often included in the mathematical model.

**Advantages**:

Maneuvering simulators deliver the flexibility necessary to investigate the different aspects of the real world that concern the ship control. Moreover, they provide the ability to operate in real time and allow accurate use of the human beings. Another main advantage is the ease of rerunning the program for different set of variables (changing rudder angles, RPM, engine power, type of rudder and propeller, environmental conditions), [11, 42-44].

Fast-time simulators are cost-effective and not time consuming. Fast time maneuvering trajectories can be rerun at a time fraction of real time.

**Drawbacks**:

A hydraulic simulator has the following main drawbacks. It has high cost and time requirements, and does not usually address the navigation process or accurately model the vessel trajectories because of the small ship models involved and the time-scaling difficulties in using moving models.

The case of computer simulators has also some drawbacks when are used either for research either for training.

Two modes of operations are usually practiced: fast-time and real-time. Even though only parts of a voyage usually need to be simulated, real-time operations are time consuming and costly.

**Comments**:

The approach is mainly limited only by the accuracy of its mathematical models of· ship and waterway (including current), and by the physical and visual resemblance of the arrangements to reality.

Historically, maneuvering simulators have been associated with training instead of research. However since the 1970's research oriented maneuvering simulators has been widely used to investigate problems associated with marine transportation. Today maneuvering simulators, particularly the ones using computer simulation, are largely applied in research, design, development and training.

### 3.3. Full Scale Trials

**Goal**:

To evaluate ship maneuverability by means of full-scale trials and analysis, involving either few definitive maneuvers or a complete set of tests.

**Description**:

This approach is based on the measurements and analysis of dynamic variables of full-scale trials. Many different kinds of tests, such as a speed trial, a turning circle, a stopping trial, etc. can be carried out to know the maneuverability of ships. The results of this approach can be used in two ways. One is the use of the results for a full-scale assessment of the maneuverability. The other way is to employ a system identification techniques to calculate the hydrodynamic force, [45J, to be used with a maneuvering simulator. Because of the variety and methodology of trials, some organizations have developed maneuvering trials codes. At least four national and international bodies have such codes, however these codes do not agree in details. Table 1 shows the proposed coverage of maneuvering trials codes of British Maritime Technology (BMT), Society of Naval Architects and Marine Engineers (SNAME), Det Norske Veritas (DnV), and the International Towing Tank Conference (ITTC).

All four organizations agree on the need to conduct three specific tests:

• Turning maneuvers from full speed
• Zig-zag or Z maneuvers
• Crash stop from full speed

A description of maneuvering characteristics and methodology of full-scale trials is included in [8].

Table 1
Maneuvering Trial Codes

| Trial | BMT | SNAME | DnV | ITTC |
|---|---|---|---|---|
| Crash stop ahead at full speed | * | * | * | * |
| Stopping trial at low speed | | | * | * |
| Coasting stop test | | | * | |
| Crash stop astern | | * | | |
| Stopping by use of rudder | | | * | |
| Turning test at full speed | * | * | * | * |
| Turning test at medium speed | | | | * |
| Turning test at slow speed | * | | * | * |
| Turning test with propulsion stopped | | | * | |
| Turning test from zero speed | * | | | * |
| Pullout (from turn) | * | | | * |
| Weave manoeuvre | * | | | |
| Z manoeuvre | * | * | * | * |
| Direct spiral | * | | | * |
| Reverse spiral | * | | * | * |
| Statistical method | * | | | |
| Change of heading | | | | * |
| Lateral thruster: | | | | |
|     Turning Test | | | * | * |
|     Course-keeping test astern | | | * | |

However, at least the following characteristics can be estimated in full-scale trial conditions:

• Turning circle characteristics. These can be determined from the turning circle tests using a rudder angle of 35 degrees.

• Yaw checking capability. This can be determined by the first overshoot angle and the time to check the yaw in a zig-zag test maneuver.

• Initial turning capability. This can be determined at the beginning of the zig-zag maneuver from the change of ship's heading angle per unit rudder angle and the distances traveled ahead and to the side after a rudder command is executed.

• Course-keeping capability. While no single measure of course-keeping ability has yet been developed, this quality may be evaluated for ships of comparable type, size, and speed by comparison of Z-maneuvers, direct or
reverse spiral tests, and pullout (from turn) tests.

• Slow steaming capability. The ability to proceed at the steady slow speed. It is usually determined only from the ship's speed associated with the lowest possible engine RPM.

• Stopping capability. This can be determined from the distance the ship travels along its track once a crash astern order is given. Turning while stopping from higher speeds is also of concern, and should be measured and recorded.

**Advantages**:
The full-scale trials provide a method of verification of maneuvering performance in given environmental conditions. Full-scale representation of turning circle, stopping and other maneuvers can help ship operators to understand the maneuvering behavior of the ship.
Moreover, system identification results can be used to design simulators for specific training or operational research.

**Drawbacks**:
There are at least two main drawbacks. Firstly, the approach is costly and time consuming. Although, full-scale trials are often included in the contract and usually performed at the acceptance trials, only the minimum runs of maneuvering tests are done, e.g. turning circle and stopping trial. It is not usual to run any other additional trial. So, the maneuvering assessment can be incomplete. In addition, the data gathered is applicable to similar ships in identical

environmental conditions, but it is not helpful to let it provide some general guide to new ship maneuverability.

**Comments**:
Ship maneuverability continues to be assessed by means of full - scale trials, involving either few definite maneuvers or a complete set of tests. It is still the most reliable method for understanding the maneuvering performance of a ship. The approach has been in use for many years, so that considerable experience now exists.

Ship handlers often appeal to full-scale analysis to develop an understanding of a ship's maneuvering performance. Ship designers can benefit of full-scale trials of existing ships when designing similar ships. On the other hand, the approach is helpless when developing new type of ships. Finally, the approach facilitates to correct possible maneuvering deficiencies before ongoing commercial navigation.

### 3.4. Model Experiments

**Goal**:
To provide experimental evidence in support of theoretical prediction for ships where maneuvering capability is very important or for research purpose.

**Description**:
Model experiment techniques used in model testing are classified in the following sub-headings:

• Free running model tests. The technique makes use of a self-propelled scale model of the ship fitted with all appendages and remote control, so that actual maneuvers can be performed and maneuverability evaluated.

• Captive model tests. The technique makes use of a model towed along a tank or it is towed round a circular path, or the model is oscillated while it is being towed along a tank. Then, some known motions are imparted to the model and the corresponding fluid forces are measured with some form of dynamometer.

Generally, free running model testing can use the same set of trials as at full-scale.

The captive model testing aim is to measure the hydrodynamic force and moments acting on the hull and appendages when the vessel is given a specific motion. By conducting a series of tests, changing only one parameter of motion at a time it is possible to evaluate how the forces and moments change with motion parameters.

Analyzing the results, the hydrodynamic coefficients for use in the equations of motion can be calculated.

The simplest captive model test is the OBLIQUE TOW TEST, [8, 37, 38]. It can be conducted without special facilities in a normal towing tank. Basically, the test consists of towing the model down at constant speed similar to a resistance test. The run is repeated for different drift angles to the direction of the motion. The test yields sway and yaw moment linear coefficients.

In the ROTATING ARM TEST, [8, 37], the model is supported on a carriage which can be moved along the arm to vary the curvature of the path. Also the yaw angle can be changed. Series of tests can be conducted with various combinations of curvature of path and yaw angle. Also the speed of the arm can be controlled. The arm is designed to accelerate to constant speed. With this, linear and nonlinear coefficients can be calculated.

Another technique is by means of the PLANAR MOTION MECHANISM (PMM), [8, 21, 37]. This device can yield linear coefficients and non-linear ones if the motion is made large enough. The principle of PMM is that the model is forced to oscillate whilst being towed down a ship tank.

The resulting motion is a sinusoidal yawing and swaying.

The normal method of testing is to conduct two separate forms of oscillatory motion namely pure sway and pure yaw. On the other hand, combined motion tests can be al so conducted.

The Oblique Tow, Rotating Arm and PMM are complementary to one another. Complete information on the hydrodynamic forces is obtained by combining the results of the three facilities.

**Advantages**:
Free running model tests have at least three advantages:
• direct evaluation of performance of specific design;
• can be economical and expedient;
• can be used to study maneuvers which may be too complicated to examine by analytical or computer methods.

Captive model tests have at least the following advantages:
• provides basic data on which to support design of individual body and appendages;
• provides basic data which can be used with the equations of motion in computer and simulator studies to evaluate inherent and close-loop performance;
• provides data which can be directly utilized in design of automatic control systems;
• results can be utilized to study effects of proposed design changes without the need for additional model tests;
• data are perpetuated so that further studies can be made at a later time without the need for additional model tests;
• provides a powerful tool for doing research or systematic series work;
• dynamic scaling not necessary.

**Drawbacks**:
Free running model tests have at least the following drawbacks:
• does not provide data which can be directly related to the design of body and individual appendages;
• does not provide data to directly support design changes to effect improvements in performance;
• Froude or dynamic scaling necessary for mass, moments of inertia, and metacentric stability.

Captive model tests have the following drawbacks:
• data are one step removed from directly indicating all the handling qualities;
• method may become uneconomical for evaluating performance of one specific design;
• requires expensive facilities.

In addition, experimental results have some other limitations including an undetermined maneuvering scale effect.

**Comments**:
This approach provides a practical design method for assessing a prototype performance. The cost and time impacts of the tests on a design effort are of course part of any testing, and realistically only the final design configuration or 1 or 2 variants can be studied.
Correlation of the scale model results to full scale is not yet fully investigated and may generate problems when investigating some maneuvering performance using free running models. However, the results of the experimental approach are useful when investigating new type of ships or control devices.

It is definitely one of the most popular ways of studying maneuverability at a detailed design stage.

## 3.5. Statistical Analysis

**Goal**:
To assess maneuvering performances on the basis of previous work on similar hull forms whether full scale or model scale.

**Description**:
Maneuverability of similar ships evaluated from the results of full-scale or model-scale trials for many years is collected. Afterwards, the results are statistically summed up to provide the advance, the transfer, the tactical diameter, the steady turning radius of turning circles, and other major maneuvering parameters. When historical data from similar ship tests are accurate enough, it is possible to predict the maneuverability of the planning ship. Generally, a

statistical method makes use of two approaches. One is to use trial's data of similar ships to assess maneuvering performances such as stopping distance, turning diameter, etc., [12]. The other is to estimate the hydrodynamic force on the basis of previous hydrodynamic data measured on ship or model tests, [9, 41]. Once the coefficients are known a computer simulation can show the predicted maneuvering performances of the planning ship.

Sometimes, some maneuvering qualities of a ship design can be quickly predicted by using specific charts for determining rudder area, which are based on the results from a number of maneuverability tests with actual ships.

The principle of this method is again to judge the follow-up maneuverability by comparing with actual results from conventional ships.

**Advantages**:
This approach is another popular, most promising and inexpensive procedure available for predicting maneuvering performance, particularly if data of similar ships are at hand. When using full-scale data it is possible to predict maneuverability without having many correlations' problems.

**Drawbacks**:
It is very difficult to take precisely into account the differences of ship types. It is rather doubtful to apply old data to the newer ship types, and special care is needed when using actual data for full-bodied ships because just a slight difference of arranging propellers or a rudder or difference in stern form may give quite different maneuverability.

New types of design cannot be investigated because of the lack of data of previous work. This approach constrains the choice of concepts for new designs.

**Comments**:
This approach needs an analysis of previous data. Ship designers cannot realistically perform such an analysis that requires specific knowledge. Research institutions are able to perform such analysis. The approach can also be combined with results of theory. The theoretical description is useful in the context of interpolation between sparse experimental data. The approach can be more cost-effective than running model experiments.

3.6. **Discussion**

On the basis of the features examined, it is useful to look to the following issues:

• Mathematical model technique. It has often been used since the wide availability of high speed computers. This technique provides a solution and basic understanding of the dynamics of submerged and floating bodies. However, the differential equations of motion are comprised of a number of coefficients or derivatives which are of hydrodynamic origin. Consequently to obtain reliable solutions for any given configuration, it is necessary to know the coefficients with a reasonable accuracy.

• Use of Maneuvering Simulation. In the past, maneuvering simulators have been used for training. Since the 1970's, research oriented maneuvering simulators are applied to investigate problems associated with marine transportation. The approach is mainly constrained by the accuracy of its mathematical models.

• Full-scale Maneuverability. It is one of the most reliable methods for understanding the maneuvering performance of a ship. The approach has been in use for many years, so that considerable experience now exists.
Ship handlers often appeal to full-scale analysis to develop an understanding of ship maneuvering performance. Ship designers can benefit of full-scale trials of existing ships when designing similar ships. On the other hand, the approach is helpless when developing new types of ships.

• Experimental Assessment. The cost and time impacts of the tests on a design effort are of course part of any testing, and realistically only the final design configuration or 1 or 2 variants can be studied.
Correlation of the scale model results to full scale may generate problems when investigating some maneuvering performance using free running models. The results of the experimental approach are useful when investigating new types of ships or control devices.

• Statistical Data Assessment. Using statistical methods, an analysis of previous data provides an assessment of similar ships. Research institutions are able to perform such analysis. The approach can be more cost-effective than running model experiments.

# 4 APPROACHES TO SHIP SAFETY

When studying safety it is necessary first of all to find an agreed definition of it. Several sources, [1-5, 17-20, 46-54], provide different understandings of safety.

Where, the dictionary supplies the following definition:

"Safety is the condition of being free from undergoing or causing hurt, injury, or loss".

While, for example, safety professionals may use the following one, [55]:

"Safety is a quality of a system that allows the system to function under predetermined conditions with an acceptable minimum of accidental loss".

Notwithstanding, in the present context, it is beneficial to initiate by stating a fresh definition of the word "Safety" as presented by Kuo, [26]:

"Safety is a perceived quality that determines to what extent the management, engineering and operation of a system is free of danger to life, property and environment."

There are a number of words in this statement which call for examination:

Quality: Safety is not something absolute, but a quality to be specified according to given circumstances and can be continuously enhanced over a period of time as a result of new decisions and increased experience.

Perceived: Safety is perceived quality because it depends on actual circumstances and the competence and experience of individuals involved in a situation.

System: the word is used to represent any complete installation, i.e. a ship, or a component of an installation, process or project.

Management: A System is introduced to meet a specific objective, which is implemented by the management of the organization.

Engineering: The role of engineering or technology in safety is best understood by engineers because it involves technical aspects.

Operation: Operation is important because even the most carefully thought-out system could fail trough incorrect operation. It is also virtually impossible to cater effectively for the interaction of all possibilities, especially in the case of more complex systems such as ships.

Life: Safety is most closely associated with human life, and that is the right emphasis. In practice, a human being is exposed to many different types of danger in different activities, and it would be impossible to achieve absolute and total safety in this respect.

<u>Property</u>: The term "property" covers both the system of interest and other systems may be endangered by it in any way.

<u>Environment</u>: Marine failure and others as well, can affect the environment in a most significant way.

The study of ship safety is basically directed in three directions. One is the development of safety rules to be satisfied by the system, such approach is called "Prescriptive". Another one is the application of risk analysis techniques to deal with the technical safety aspects of the system. The other one is to approach safety with preventive methodologies.

This chapter is divided in three parts. Each part describes one of the three main approaches to safety:

• Prescriptive;
• Risk Analysis;
• Preventive.

## 4.1. Prescriptive Approach.

**Goal**:
To provide technical consistency between systems operating in the same industry.

**Description**:
The approach makes use of past practical and scientific experience within findings of major disaster to develop and eventually improve the safety requirements. The requirements are then translated in rules which are incorporated, verified and controlled during the entire life-cycle of the system.

There are rules covering structural aspects as long as dynamics and operational aspects as the system. Typical examples of such approach are the rules, [56-59], used by the International Maritime Organization (IMO) which is the United Nation Agency for Maritime Affairs, by National Governments or by Classification Societies such as Lloyd's Register, RINA (Registro Italiano NAvale), Det Norske Veritas.

**Advantages**:
This approach incorporates the past experience in a new system, particularly when involving ships similar to those already in service, [53-55]. Moreover, action following major accidents will generate new or improved rules so that new designs will benefit from it.

**Drawbacks**:
Rules and regulations are usually behind the latest technological advances or practice. This approach is also a passive way in which safety is incorporated into the design. The use of major disasters for stringent requirements can make possible that too much emphasis is put on the causes of one event and this may mislead the prevention of other unforeseen problems.

**Comments**:
The thinking behind this approach is that rules and regulations can reflect past experience. The approach is compulsory in practical ship design. It is a popular way of respecting safety requirements, [19]. However, its usefulness is limited, especially in new situation.

## 4.2. Risk Analysis

**Goal**:
To decide whether a potential hazard has to be considered for further investigation on the basis of associated risks.

**Description**:
Risk analysis considers mainly two aspects of accidents: frequency and severity, [46,52]. Frequency refers to the likelihood of occurrence. Severity refers to the impact of loss in terms of money or physical impairment. Using these factors, a risk analysis develops classes of severity and frequency. These classes are used to rank the relative risk of various events, see [53, 55]. Each class is defined in terms meaningful to the organization.

For example, "catastrophic" may, to one organization, refer to a defined value of losses and which involves death or dismemberment, while another organization may have lower or higher standards. Events are ranked in terms of probability and severity, and the ones that fall in the categories of combined risk are given priority.

Applying risk analysis to identify potential hazards, it allows their level or risk to be estimated and activates the decision making process to its acceptability.

**Advantages**:
Other industries have applied this approach. This know-how can be easily transferred to ships. The search for potential hazards is done in a systematic way.

**Drawbacks**:
It is difficult to define common classes of frequency and severity. The approach relies mainly on past data available on the probability of occurrence of failure.

Moreover, mis-prioritising potential hazards may be counter-productive or even lead to more serious hazards.

**Comments**:
Application of Risk Analysis may reveal aspects of the system which requires more consideration. Results of the analysis are used for the judgment about the acceptability of the risk and for decision making.

However, in practice there is a tendency to place too much weight on numerical results, which can be highly misleading. This is because, even when based on accurate data, the derived results will only provide information regarding technical aspects of the ship. These aspects may not necessarily be the most significant in comparison with human factors and operational procedures.

### 4.3. Preventive Approach

**Goal**:
To assess and control the total safety of a system over the ship's entire life-cycle.

**Description**:
In one of his paper, [26], Kuo refers to safety aim as:

"to anticipate potential or expected failures by taking active steps to minimize the likelihood of their occurrence or to limit their effects."

This definition well explains the contents of the preventive approach. It is al so useful to remember that the approach generates from the fact that when a situation exists that creates loss or injury and when something can be done to avoid it, it is more effective to prevent the event to happen. Preventive approach is largely applied in medicine. Other industries have examples dating since the 50's. However none of the approaches is so comprehensive of the total safety like the one suggested by Kuo, [26, 60-62]. This approach consists of covering all the aspects of safety including the previous two approaches (4.1 and 4.2) and still being cost-effective. Kuo suggested 9 main steps to implement safety in a system, [26]. They include the identification of potential hazards, the study of causes and consequences, the involvement of human actions, the scope for design modification, the engineering of containment system, the use of viable emergency solution and sound practice; finally the approach checks the interface with rules and regulations and includes the training of personnel. The approach is applied to each single phase of the life cycle.

**Advantages**:

The preventive approach is a "total safety" approach very comprehensive of the different elements concurring to make safe systems. It all so satisfies both operational requirements and the demand for cast effectiveness and high level of safety. It can be applied during the entire life cycle of a system.

**Drawbacks**:

Due to the particular type of method, it could be said that there are not drawbacks to this approach. Its application requires an involvement and an inside understanding of the entire life cycle of a ship, including both management and engineering. For these reasons it could be argued that its application may become more tedious than other approaches for ship designers and builders. On the other hand, a better understanding of the system will generate a safer and more cost-effective engineering system.

**Comments**:

Ships and marine vehicles are becoming more complex with the introduction of completely new designs and the use of advanced technology. The preventive approach can keep pace with such advances while coping with rules and management. Finally, drawbacks far designers and builders will turn to advantages for customers.

## 4.4. **Discussion**

There are a number of other approaches directly aimed to improving system safety. From the design point of view, each of the approaches, including the above ones, has a contribution to make to some aspect of safety. However, most of them are unlikely to satisfy the desired goal or the requirements of the safety definition.

On the basis of the approaches discussed the following issues look far examination:

• Statutory or Prescriptive Rules. The philosophy of this approach is that rules and regulations can reflect past experience. The rules are compulsory in practical ship design. However, its usefulness is limited, especially in new situation such as nonconventional designs or innovative technology.

• Reliability or Risk Analysis. The application of Risk Analysis usually reveals some aspects of the design or system which requires more consideration. In practice there is a tendency to place too much weight on numerical results, which can be mis-leading. This is because, even when based on accurate data, the derived results will only provide information regarding technical aspects of the ship. These aspects may not necessarily be the most significant in comparison for example with human factors and operational procedures.

• Preventive Approaches. Ships are becoming more complex with the introduction of unconventional design and the use of innovative and advanced technology. This approach can keep pace with such advances while coping with rules and management. The PREVENT-IT safety methodology can offer an effective solution for devising safer ships.

# 5 ATTEMPTS TO RELATE SAFETY WITH MANOEUVRABILITY

In the past, there has been a vague idea of what relating safety with maneuverability means. Moreover, the regulatory bodies-say IMO (International Maritime Organization), classification societies and national authorities-have not yet provided a common understanding of standards or a common explanation which can define the maneuvering safety property adequately. However, there are at least four international bodies which have some interest in the subject. The IMO has a standing committee on ship maneuverability and has produced and is still developing a set of guidelines and standards for design and operation, [19, 25]. Another one is the Panel H-10 (Ship Controllability) of SNAME (The Society of Naval Architects and Marine Engineers, U.S.A.) that is working in this area and has produced systematic studies of maneuverability and is producing a maneuvering design workbook, [22]. The panel is also very active in setting maneuvering standards working together with the US Coast Guard and US Pilot Associations, [18, 27]. The third one is SNAJ (The Society of Naval Architects of Japan) which has a Technical Committee with the purpose of improving the prediction of ship maneuverability; they have produced analytical methods and technical recommendations, [15]. While, Lloyd's Register of Shipping has issued in 1986 the "Provisional Rules for the Classification of Ships Maneuvering Capability", [59J.

Other bodies have suggested attempts to relate safety with maneuverability looking to a specific aspect of the problem, say operational procedures, or constraining restricted area to ships with specific maneuverability.

The whole subject can be divided in four main approaches:

- the approach which makes use of design guidelines;
- the approach which looks to the operational side requiring <u>maneuvering information be provided to the on-board operators</u>;
- the tentative of setting up <u>maneuvering safety standards</u> at full-scale;
- the use of auxiliary or <u>supplementary maneuvering controls</u> to improve poor ship maneuvering performances.

The aim of this chapter is to review these approaches.
The chapter is divided in four parts. Each part describes one of the four approaches.

## 5.1. Design Guidelines

**Goal**:
To provide criteria for estimating maneuvering qualities to enhance safety at the ship design stage.

**Description**:
The approach looks as a safety prescriptive approach, but it does not have the compelled mandate of other regulations. It is based on the IMO Resolution MSC/Circ. 389, 1985 "Interim Guidelines for Estimating Maneuvering Performance in Ship Design", [13,19,58]. The guidelines declares that, in order to improve operational safety, all ships should have satisfactory maneuvering qualities which permit them to:

- keep course
- turn
- check turns
- operate at acceptable low speed
- stop

This guideline is applicable to all new ships greater than 100 meters in length. The resolution requires that, at the design stage, the maneuvering performance, similar to the one at full-scale, see section 3.3., may be estimated for a particular ship by calculation, model testing, or use of data from similar ships (statistical approach). It is also requested that full-scale tests are carried out on the ship and then compared with the design estimations, in order to refine the whole estimation process. The full-scale trials required are mostly based upon the ITTC code, see section 3.3.

**Advantages**:

The approach activates a safety design procedure regarding maneuverability. If national authorities enforce it, the guidelines force the ship-builder to consider maneuverability during the design stage and to indicate whether they have done so, and to give an explanation of the methods which have been used.

**Drawbacks**:

The guidelines seem too vague to be really employed and understood by ship designers. In any case it is a passive way in which safety is incorporated with maneuverability.

**Comments**:

The approach recognizes that most maneuvering qualities are inherent of the design and they should be evaluated during the design process In fact, if at the design stage the ship inherent maneuverability is predicted, ship designers can possibly avoid designs with "poor" maneuverability. However, to be effective the "guidelines" have to be enforced by National Administrations and then applied to all new ship designs as required. Even though the guidelines provide that maneuvering evaluation is important at the design stage, it is not completely clear in which way maneuvering-safety is implemented at the design stage. The only real outcome of setting this guideline is the fact of attracting some attention on the subject.

## 5.2. On Board Maneuvering Information

**Goal**:

To enhance operational safety providing full-scale maneuvering information to the ship operators.

**Description**:

This approach is a typical prescriptive safety approach, its application is compulsory far some types of ships.

It is based on a 1987 IMO Resolution A.691 (XV) "Provision and Display of Maneuvering Information On Board Ships", [13, 19, 56]. Safety is achieved providing maneuvering information to the operators. The resolution declares that on board the following information should be available:

- Pilot Card
- Wheelhouse Poster
- Maneuvering Booklet

The National Authorities should recommend that the following information be provided:

- The Pilot Card, for all new ships to which the requirements of SOLAS (Safety Of Life At Sea) IMO International Convention of 1974 apply;

- All three, for all new ships more than 100 meters of length;

• All three, for all new chemical tankers and gas carriers regardless of their size.

The Pilot Card contains: ship dimension and loading condition, propulsion and maneuvering equipment, and other relevant navigational equipment.

The Wheelhouse Poster, to be permanently displayed in the wheelhouse, contains the general particulars of the ship and its propulsion machinery. It contains: turning circles and stopping maneuvers, with loaded condition on the left handed side and the ballast condition on the right hand side. A man overboard maneuver should also be shown.

The Maneuvering Booklet contains comprehensive details of the ship's maneuvering characteristics, which should include both the information shown on the Wheelhouse Poster and Pilot Card, together with other available information.

**Advantages**:
On board operators such as navigators and pilots can have an understanding of the ship maneuverability in given environmental conditions and then, based on their own expertise, to extrapolate in various navigational situations.

**Drawbacks**:
It does not improve human performance. It does not implement any maneuvering requirements. It is a gross "over-simplification" of such very complex problem. As a result, on board operators can take decisions based on erroneous assumptions because the information is valid only for a specific set of environmental conditions.

**Comments**:
The human involvement is approached the wrong way. The objective of providing information is not accomplished since only some valid and scattered results of maneuvering trials are provided. So also the point of providing information can result improved safety standards is not met. Moreover, the communication style of the information is not positive, because it is unlikely for operators to have time to study this information during maneuvering operations.

## 5.3. Maneuverability Standards

**Goal**:
To improve operational safety requiring maneuvering consistency between full scale ships.

**Description**:
This approach is a classical safety prescriptive approach applied to the maneuverability case. Maneuvering criteria for conventional ships have been proposed through the years as research on maneuvering has progressed. In 1944, Kempf was the first to propose criteria, [23]. He proposed executing a 10 degrees zig-zag test with the rudder angle switched four times, the distance from the first change to the point where the heading is back to the original course divided by the ship length is to be between 6-10 for "reasonable qualities". Since then, several criteria have been researched and proposed by different authors, [8, 12, 18, 24, 27, 29, 63-70].

IMO has developed Ship Maneuverability Standards, [25], that are likely to include:
• course keeping ability, based on 10 degrees zig-zag test and first and second overshoot angle;

• stopping ability, track < 15 L (Ship Length).

During the testing the ship speed should be 90% of the full speed or more. The proposed standards are only for full-scale ships.

**Advantages**:

The standards give an explanation to the definition of "good" or "satisfactory" maneuverability. It is a clear attempt to give a definition and a measure to acceptable maneuverability. If such standards are approved and activated, the ship design stage will be forced to take in account such requirements. New designs will benefit directly from their application.

**Drawbacks**:

The proposed standards are not fully comprehensive of the whole range of maneuvering performance. Their application can overlook some other relevant maneuvering aspects that can lead to serious casualties. Moreover, the standard regulation can be a long way behind the latest engineering practice.

**Comments**:

Maneuvering standards support the idea that ship maneuverability is itself a safety requirement. If ships can satisfy a common maneuvering consistency the human involvement will be decreased. It also gives a method for understanding what is an acceptable level of maneuverability that can be useful to ship designers, owners and operators.

### 5.4. Adding Maneuvering Appendages

**Goal**:
To improve ship operational safety by adding specific maneuvering appendages.

**Description**:
Even though it cannot be regarded as an exemplary safety approach, many organizations apply it to improve maneuvering-safety. Maneuvering-safety is not considered important until evidence, either at full or model scale, shows that poor maneuverability will affect the ship safety and operability and so its economic benefits. In this approach a corrective action is considered when the operational safety of the ship is in jeopardy. Generally, there are two ways to acting. One is to act at the design stage. The hull is chosen based on other requirements and then the maneuvering requirements are satisfied adding the appendages. The other way is to improve or change the maneuverability of seagoing ships to meet the requirements such as new standards or routes or because of "poor" maneuvering performances. Several active controls are used. Typical examples are: rudders, propellers, lateral thrusters, ducted propeller, others, or combinations of them. Specific details of such appendages can be found in [30].

**Advantages**:
Inadequate maneuvering performance can readily be improved with specific maneuvering appendages.

**Drawbacks**:
The adding of specific and complex arrangements of maneuvering appendages can generate difficulty far on board operators. They can be mis-leaded to accidents because of the wrong use of combination of the appendages. Moreover, the method is not cost-effective because of the high cost of such appendages and their installation.

**Comments**:
In some cases, other stringent requirements may not make cost-effective to consider maneuvering specifications.

The method may be the only plausible solution to achieve maneuvering and safety requirements. It is a popular way of enhancing maneuverability. It is not unlikely to find ships with complex arrangements of lateral thrusters, or other maneuvering devices.

# 6 SCOPES FOR ALTERNATIVE METHOD

## 6.1. Practical Exigencies

Some investigations of recent major rammings and collisions, [5, 6], by the Marine Accident Investigation Branch of the Department of Transportation of the United Kingdom and the National Transportation Safety Board of the U.S.A., concluded that the causes of most of those accidents can be summarized in three groups:

• <u>inherent poor maneuverability of the ship</u>;

• <u>human errors in misjudging or misunderstanding the maneuvering behavior of the ship</u>;

• <u>faulty maneuvering systems</u>.

In some cases, the investigators reported that the causes were somehow inter-related. These conclusions support the fact that maneuverability has to be considered over a ship's life cycle. Indeed, the two following considerations can be readily done:

• inherent maneuverability is definitely established during the initial phases of the ship's life cycle, see Appendix 6 and [9, 22, 28, 30, 63], normally known as the design stage. On the other hand, the other phases are still vital to the inherent maneuverability. For example, if the designer has well considered and implemented the required maneuverability, a lack of quality control in the other phases – say procurement and construction – can invalidate the efforts of the ship designer.
• human errors in misjudging or misunderstanding the maneuvering capability of the own ship are likely to be related to the operation phase of the ship's life-cycle, see [1, 2, 4, 71]. However, misinterpretation of the role of the human involvement and the training of the on-board personnel during the other phases can lead to such errors. Some studies report that approximately 95% of maneuvering accidents are due to human error, [72].

The maritime industry is awaiting and hoping for more findings from the researchers and developers to rightly face the problem of generating better maneuvering and safer ships. The problem becomes peculiar and more complex when evolving a ship destined for unusual services and/or requiring specific superior maneuvering characteristics. On the other hand, one may ask if this is really a problem at all far many ships of the type and function whose maneuverability and safety have been shown to be "acceptable", see Appendix 1 and [63, 64], aver the years. The opinion here is that, even far these vessels, it would be beneficial to know that once the principal characteristics, form and active controls are selected the design would lead to a ship of improved maneuvering-safety characteristics that most of the past, within other constrains. So that the customers, i.e. owners, operators and passengers, will have a higher maneuvering and consequently safer ship far the future they have to live with the design. This includes the entire life-cycle of the ship.

## 6.2. Improving Productiveness and Safety

The benefit of a life-cycle approach is that a highly maneuverable ship may well be able to cut down docking fees, improves its safety by being better able to avoid a collision, and even possibly to receive preferential treatment by harbor pilots and avoid delays required by the pilot to adjust to the ship or trim conditions, wait far more favorable tides or current and in docking when tugs are not available. Furthermore, the benefits are more important far ships whose service requires higher maneuverability.

Moreover, it is necessary to underline that the maritime industry has an international and cyclic nature. It has been argued that for success in the maritime business environment an organization must have the right market, management, money and manpower, and also ensure that these four Ms are developed in a balanced way at all times, see [7]. Indeed, the word "management" is also part of the safety definition. Management is interested in profits and the return on investments. If management can be convinced that safety and maneuverability implementation is cost-effective, i.e., that they will reduce operating expenses and increase profit, it is more likely to support the safety-maneuvering program. To convince management, ship designers must support the fact that maneuvering-safety is in the company's best interests as well as employee's. So when facing the problem of maneuverability and safety, it should be accentuated that the maritime industry needs a philosophical way of thinking which will keep peace for all the requirements of its business activities.

### 6.3. Weakness of Previous Approaches

Scattered maneuvering-safety approaches are not the solution to the problem. Unfortunately, in the previous approaches, there are some ideas of guiding the maneuvering assessment during the design process and at full-scale or to provide information to the personnel, which are just a part of the entire life-cycle. There is not clear interface between safety and maneuverability.

For example, the proposed maneuvering standards will be based on full scale trials at the commissioning phase of a ship's life cycle. However, their application will possible enhance one of the other approaches such as using the design guidelines, or adding maneuvering appendages which may belong to some other phases.

Moreover, it is important to underline some basic considerations which arise from some unconvincing points of previous attempts to relating safety with maneuverability and that will be useful for devising a sound methodology with wide utility:

• <u>Simultaneously satisfying operational and safety requirements</u>. Different operational requirements, including maneuverability as well as safety ones, have to be concurrently satisfied at the design stage;

• <u>Providing for technological advances</u>. Innovative technology may influence or mis-lead the understanding of the concept of "acceptable maneuverability". Moreover, the enforcement of maneuvering-safety standards may diminish the development of innovational concepts. This should be taken into consideration when integrating safety with maneuverability.

Finally, the main weakness of the previous attempts is the lack of a methodology to relate safety with maneuverability during the entire ship's life-cycle while satisfying the demand for other operational specifications and for cost-effectiveness. However, to be effective, any methodology will firstly need to provide an understanding of "good" or "acceptable" maneuvering qualities, this is contributed in the following chapter. On the other hand, the problem of deciding how to discriminate between ships with acceptable and unacceptable maneuvering-safety features needs a proper solution. A suitable solution can be achieved incorporating safety features with ship maneuvering qualities at the appropriate phases of a ship's life cycle-say ship's design stages-to later ones.

Such a solution is discussed in the following chapters.

# 7 BASIS OF ALTERNATIVE MANOEUVRING CRITERIA

## 7.1. Scope for Alternative Indices of Maneuverability

The need of deciding what the acceptable level of Maneuvering Qualities is has stimulated the development of a variety of quantitative measures of ship performances based on full-scale trials such as turning circle or zig-zag maneuvers. Steady state amplitudes, geometric features of steady trajectories, and transient state exponents (e.g. Nomoto indices, [65]); have been employed as maneuvering indices since the mid-fifties, [8, 12, 19, 65-68]. More recently, overshoot angle, time to overshoot, side and head reach, time-to-second-execute, P-ratio and phase advance have been used to assess the dynamic stability as well as the maneuvering capabilities of ships, [12, 67-70].

Although, these measures have been largely discussed and used and some of them are being used to set the IMO Maneuvering standards, [25, 27], they do not be completely represent the ship-handlers' requirements, [17, 68, 73].

Despite the available measures of maneuverability assess the radius, distance, angle and time required to accomplish a maneuver and they can be used to assure adequate ship maneuverability, the ship-handler's interest centers on the transient effects of small course corrections during a maneuver and on the ease with which the maneuvering changes can be accomplished, [17, 68, 73].

Moreover, the conventional methods to assess those maneuvering qualities require full-scale trials such as spiral test, reverse spiral test and zig-zag maneuvers, [8, 15, 29, 67]. These testing techniques can prove to be expensive and time consuming. Therefore, there is scope far investigating alternative ways of measuring ship's maneuverability that satisfy the ship-handlers' requirements.

However, before suggesting alternative ways of measuring ship's maneuverability, it is necessary to understand what are the ship-handlers' maneuvering qualities' requirements. According to previous studies, [16, 68, 73], at least the following maneuvering qualities are of importance in piloting situations:

• the ship ability to promptly response to the helm from a straight course, i.e. the responsiveness to rudder; and

• the ship's ability to turn onto the straight steady course once the rudder is returned in the neutral position when in a turning motion, i.e. the straight line stability.

Besides these requirements, it has also been argued, [17, 73] that from the ship-handler's point of view, the most advantageous situation during a maneuver is that ideally the sway speed v should decay exponentially to zero, while the yaw rate r should decay at the slower rate, [17, 73]. Thus the ship can turn onto a new course with essentially no drift.

The aim of this chapter is threefold:

• to introduce the basic properties of alternative indices of maneuverability from the ship-handler's point of view;

• to analyze ship's maneuvering performances using the alternative indices; and

• to propose maneuvering criteria based on the alternative indices of maneuverability.

Following, the alternative maneuvering indices, their properties and the associated criteria are derived and analyzed. Then an example is presented for illustrating the physical meaning of proposed indices. Appendix 1 illustrates the classification rules derived from the proposed alternative criteria.

### 7.2. Basis of Alternative Maneuvering Indices

During transit, it is useful to introduce a factor called "Q" which is the rate at which the sway speed v varies with respect to yaw rate r:

$$Q = \frac{dv}{dr} = \frac{dv}{dt}\frac{dt}{dr} = \dot{v}/\dot{r} \qquad (7.1)$$

This rate is an indicator of the ship's tendency to alter its drift as it turns. For example, on a straight course the ship's sway speed v is zero. From a ship-handler's point of view, the sway speed v would ideally be kept to zero as the ship is turned to a new course.

In such a case Q would be zero. On the other hand, such value is unrealistic because ships tend to develop sway speed during a maneuver. However, two ships with different values of Q will have different tendency to alter their drift during a maneuver. The ship with the smaller value of Q develops, during a maneuver, a smaller drift than the ship with a bigger value of Q.

Thus Q can be used to describe the main ship-handlers' interest in the transient effect of small course correction.

Figure 2 represents two v-r sample phase diagrams. Phase diagrams are referred as response curves describing the approach of a bare hull to the steady straight state r'=0 and v'=0. To obtain the diagram, the ship is first made to turn with some rate of turn and then the rudder (or the yaw stimulation) is returned to the neutral position. Geometrically, Q can be interpreted as the slope of the v-r phase diagram. The index is zero along horizontal portions of the response curve; it is nearly zero along the portions where v varies slowly with r, and it deviates significantly from zero along portions where v increases or decreases rapidly. The index is infinite at points where the curve has a vertical tangent.

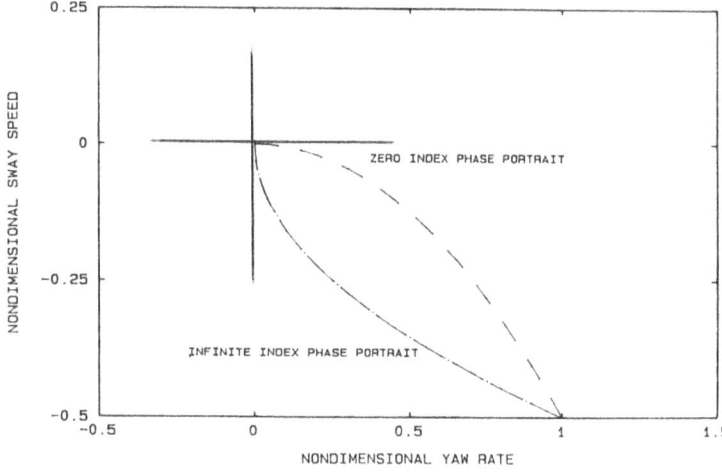

Fig. 2

Sample phase diagrams depicting the approach of a bare hull to the steady state r'=0; v'=0; the dotted path has a vertical tangent at $(0, 0)$ and corresponds to a hull with $Q_H'$ undefined. The dashed path has a horizontal tangent at $(0, 0)$ and corresponds to a hull with $Q_H'=0$.

MANOEUVRABILITY AND SAFETY OF SHIPS

Physically, Q is a ratio of the lateral acceleration $dv/dt$ ($\dot{v}$) to the rotational acceleration $dr/dt$ ($\dot{r}$):

$$Q = \dot{v}/\dot{r} \qquad (7.2)$$

Thus $|Q| < 1$ for ships that support a lateral acceleration smaller in magnitude than $|\dot{r}|$. While $|Q| > 1$ for ships that support lateral acceleration bigger in magnitude than $|\dot{r}|$.

For a ship of length L,

$$Q' = \frac{Q}{L} = \frac{1}{L}\frac{dv}{dr} \qquad (7.3)$$

is a non-dimensional index measuring the sensitivity to variations in yaw rate of the ship's sway-to-length ratio. During a steady straight course both v and r are zero.

Consequently, the right-hand side of (7.3) is indeterminate and a limiting process must be used to evaluate Q. The expression:

$$Q' = \lim_{t \, > \, \infty} (1/L)(\dot{v}(t)/\dot{r}(t)) \qquad (7.4)$$

is a <u>non-dimensional measure of the ship's ability to turn onto the steady straight course</u>. Here both $v(\dot{t})$ and $r(\dot{t})$ reduce to zero as t tends to infinity. If the ship is a bare-hull, then (7.4) is a non-dimensional indicator of the hull's inherent tendency to drift during transit to the steady straight course. In this case, Q' is referred to as bare-hull index and is denoted by $Q_H'$.

If the ship is a hull and rudder, then (7.4) defines the hull-rudder index $Q_{HR}'$. The rudder, acting as a fin, modifies the hull's tendency to drift as it turns. $Q_{HR}'$ provides a non-dimensional measure of the modified hull's tendency to sway. These indexes can be used to represent a ship's straight line stability.

While, the expression:

$$Q_R' = \lim_{t \, > \, 0^+} (1/L)(\dot{v}(t)/\dot{r}(t)) \qquad (7.5)$$

is a <u>non-dimensional measure of the rudder's ability to turn the ship from a straight steady course (responsiveness to rudder)</u>. Here it is assumed that the ship's course was initially straight and steady so that both $\dot{v}(t)$ and $\dot{r}(t)$ reduce to zero at t=0.

### 7.3. **Performances on Approach to a Straight Course.**

<u>Bare-Hull Performance</u>. Equation (7.4) provides a dimensional measure of the hull's tendency to drift as it turns onto a straight steady course. The index is evaluated by constructing the solution of the equations of motion near the steady state $U = U_o$, $v = 0$, $r = 0$, with $u_o$ fixed. Linearization of the equations of motion about ($u_o$, 0, 0) decouples the first-order surge component from the first-order sway and yaw components. For a stable hull the solution of the dominant linear sway-yaw subsystem, see (3.1), is:

$$v = c1 \in 11e^{\tau_1 t} + c2 \in 12e^{\tau_2 t} \qquad (7.6a)$$

$$r = c1 \in 21e^{\tau_1 t} + c2 \in 22e^{\tau_2 t} \qquad (7.6b)$$

where the exponents $\tau_1$ and $\tau_2$ are negative and $\tau_2 > \tau_1$.

The constant $C_1$ and $C_2$ are determined from the initial conditions. The vectors $(\in_{11}, \in_{21})$ and $(\in_{12}, \in_{22})$ are the eigenvectors associated with $\tau_1$ and $\tau_2$ respectively.

From a practical standpoint, $\tau_2$ alone is usually given for surface ships, [8]. Thus it can be shown from equations (7.6) that the motion description by the $\tau_2$ term is much larger than the $\tau_2$ term after the disturbance is ended, for example the rudder is returned to a neutral position.

Equations (7.4) and (7.6) imply that $Q_H$ is a ratio of the components of the eigenvector associated with the dominant eigenvalue $\tau_2$:

$$Q_H = \frac{\in 12}{\in 22} \tag{7.7}$$

The ratio Q is the slope of the v-r phase diagram at the steady state and can be obtained directly from the equations of motion by solving the quadratic relation for the components:

$$a21q^2 + (a22 \quad a11)q \quad a12 = 0 \tag{7.8}$$

The coefficients in (7.8) are

$$a11 = TY_v^\circ + T_y N_v^\circ \tag{7.9a}$$
$$a12 = T(Y_r^\circ - (m+m_x)u_o) + T_y N_r^\circ \tag{7.9b}$$
$$a21 = TT_z Y_v^\circ + N_v^\circ \tag{7.9c}$$
$$a22 = TT_z (Y_r^\circ - (m+m_x)u_o) + N_r^\circ \tag{7.9d}$$

where

$$T = (I_z + J_z)/(m+m_y) \tag{7.10a}$$
$$T_y = (Y_r)/(m+m_y) \tag{7.10b}$$
$$T_z = (N_v)/(I_z + J_z) \tag{7.10c}$$

and $Y_v^\circ$, $Y_r^\circ$, $N_v^\circ$, $N_r^\circ$ denote partial derivatives of Y and N at the steady state.
The quadratic relation (7.8) has two roots $q_1$ and $q_2$.
The two eigenvalues $\tau_1$ and $\tau_2$ are:

$$\tau 1 = a21q1 + a22$$
$$\tau 2 = a21q2 + a22$$

The two roots of (7.8) represent the ratio of the eigenvectors of equation (7.6) :

$$q1 = \frac{\in 11}{\in 21}$$
$$q2 = \frac{\in 12}{\in 22}$$

It can be shown that $Q_H$ is equal to the root of (7.8) that minimizes $|\alpha_{21}q + \alpha_{22}|$, i.e. which minimizes the absolute eigenvalues, [64].

It fallows that in the present context and $\tau_2 < \tau_1$ so

$$Q_H = q2 = \frac{\in 12}{\in 22}$$

and that:

$$\tau_2 = a21\ Q_H + a22$$

and finally:

$$Q_H = \frac{\tau_2 \quad a22}{a21}$$

This expression shows that the index contains the eigenvalue $\tau_2$, that is a time constant, and the coefficients $\alpha_{21}$ and $\alpha_{22}$ that are determined by the hull's hydrodynamic coefficients. The usefulness of this expression is in the fact that, for given values of $\tau_2$, $Q_H$ can be used to represent class of hulls with the same time constant. It also represents that $Q_H$ is a linear function of the time constant $\tau_2$.

The expression

$$Q_{H'} = \frac{\in 12}{L \in 22} \qquad (7.11a)$$

or

$$Q_{H'} = \frac{\tau_2 \quad a22}{L\ a21} \qquad (7.11b)$$

provides a non-dimensional measure of the hull's tendency to drift at it turns onto a straight course. $Q_H'$ is the slope of the v'-r' phase diagram at the steady state.

The greater the magnitude of (7.11), the greater the sway response per length of ship for a given change in r.

Theoretical situations. There could be two theoretical situations. One situation is $\in_{12}=0$ and $\in_{22} \neq 0$, that is $Q_H' = 0$. In this case, equation (7.6) implies that v decays exponentially to zero at the rate $\tau_1$, while r decays at the slower rate $\tau_2$. Thus the sway speed is attenuated more rapidly than the yaw rate and the hull turns onto the straight steady course with essentially no drift. The other situation is $\in_{12} \neq 0$ and $\in_{22} = 0$, in this case, $Q_H'$ is undefined and the v'-r' phase diagram is tangent to the v' axis at (0, 0) (see Figure 2). Thus a small change in r' will result in a large change in v', i.e. the non-dimensional sway speed is highly sensitive to changes in the non-dimensional yaw rate.

Practical situation. From a practical point of view, hulls are designed so that neither $\in_{12}$ and $\in_{22}$ is zero. In this case the v'-r' phase diagram is approximately linear near the origin with slope $Q_H'$. An example for illustrating the physical meaning of the index is provided in section 7.6.

Hull-Rudder Performance. If the vessel is a hull and rudder, then (7.4) defines the hull-rudder index, $Q_{HR}'$. The rudder acting as a fin, modifies the hull's tendency to drift as it turns. $Q_{HR}'$ provides a non-dimensional measure of the modified hull's tendency to sway. $Q_{HR}'$ involves indeterminate limiting process and is defined by quadratic equations involving the linear hydrodynamic coefficients, similarly to $Q_H'$.

Interest of $Q_{HR}'$ at the design stage. Equations (7.9), (7.10) and (7.11) show that Q is related to the hydrodynamic coefficients and to the eigenvalue $\tau_2$.

For a sample ship with the following main geometric parameters:

L=132 m; B=20.6 m; T=4.02 m; cb=0.596; Ra=14.62 m²; with rudder at stern; for this ship, the practical value of $\tau_2$ is usually around -.015. It is possible, from (7.11b), to investigate the effects of changes in $\tau_2$ on $Q_{HR}'$:

| $\tau_2$ | $Q_{HR}'$ |
|---|---|
| -.005 | -.414 |
| -.010 | -.349 |
| -.015 | -.285 |
| -.020 | -.220 |
| -.025 | -.156 |
| -.030 | -.090 |
| -.035 | -.025 |

When $\tau_2 = \alpha_{22}$ then $Q_{HR}' = 0$;

and when $\tau_2 < \alpha_{22}$ then $Q_{HR}$ becomes a positive value and the ship is unstable. It means that when the yaw stimulation is ended the ship will not resume a straight line course.

From the previous considerations it can be considered that $Q_{HR}'$ contains two maneuvering performance information:

• Its sign represents the straight line stability; and

• Its magnitude represents the rate of change of the sway speed relative to yaw stimulation.

Ship designers are interested in how the maneuverability of the proposed design far the ship can be improved.

From a practical point of view, it is unlikely that the ship designer can change the ship mass and the associated coefficients. It is more likely that he can have some influence on the damping forces which are dominated at small drift angles by the four linear hydrodynamic coefficients $Y_v$, $Y_r$, $N_v$, $N_r$, see Appendix 2 for an explanation of these forces.

Thus, at the design stage, an investigation on the sensitivity of $Q_{HR}'$ index with respect to the damping forces would yield trends for enhancing the maneuvering qualities of the design.

The sensitivity of $Q_{HR}'$ to the linear damping coefficients can be shown by varying one coefficient at a time. For the same sample ship given in page 76, the variation of the index is as follows:

| $Y_v'$ | $Q_{HR}'$ | $N_v'$ | $Q_{HR}'$ | $Y_r'$ | $Q_{HR}'$ | $N_r$ | $Q_{HR}'$ |
|---|---|---|---|---|---|---|---|
| -2258 | -.290 | -609 | -.290 | 490 | -.290 | -291 | -.290 |
| -2823 | -.241 | -761 | -.239 | 514 | -.278 | -365 | -.398 |
| -3888 | -.195 | -913 | -.204 | 538 | -.270 | -438 | -.512 |
| -3952 | -.153 | -1065 | -.180 | 562 | -.261 | -511 | -.627 |
| -4517 | -.119 | -1218 | -.161 | 586 | -.252 | -584 | -.744 |
| -5082 | -.091 | -1370 | -.147 | 610 | -.242 | -657 | -.861 |
| value of coefficients * $10^5$ | | | | | | | |

Even though these results are for a sample ship, they can be used as an indicator on the effects of changes in the hydrodynamic damping coefficients on $Q_{HR}'$ for a stable ship with a rudder at the stern. The following table is a summary of the effects:

| linear damping coefficient | | $Q_{HR}'$ (negative) |
|---|---|---|
| Yv' | negative increasing | decrease |
| Nv' | negative increasing | decrease |
| Yr' | positive increasing | decrease |
| Nr' | negative increasing | increase |

## 7.4. **Rudder Performance on Exit from a Straight Course**.

Rudder action turns the ship from a steady straight course by inducing a yawing moment. The moment changes the yaw rate, but it also causes the ship to sway. Ideally, the vessel would experience a change in yaw rate with no change in sway. Thus the rate at which sway is produced relative to yaw rate is an important indicator of rudder performance. The initial rate at which sway changes relative to yaw rate is given by the limit $Q_R$, see (7.5). $Q_R$ is indeterminate because the ship's motion is initiated from a steady straight course with
$v(0)=r(0)=0$. The limit is evaluated by constructing a power series solution to the equations of motion near t=0. The construction requires an assumption about the way in which the rudder angle o varies with time. Here it is assumed that the rudder angle varies linearly:

$\delta(t) = \delta_1 t$

where $\delta_1 \neq 0$ is the rudder's angular velocity.

The solution of (3.1) near the straight course expressed in series is as follows:

$$u(t) = u_o + 0t + \frac{X_\delta{}^o \delta_1}{m+m_x} \frac{t^2}{2} + \ldots\ldots$$

$$v(t) = 0 + 0t + \left[ \frac{Y_\delta{}^o}{m+my} + \frac{N_\delta{}^o T_y}{I_z + J_z} \frac{1}{T_z T_y} \right] \frac{\delta_1}{1} \frac{t^2}{2} + \ldots \qquad (7.12a)$$

$$r(t) = 0 + 0t + \left[ \frac{Y_\delta{}^o T_z}{m+m_y} + \frac{N_\delta{}^o}{I_z + J_z} \frac{1}{T_z T_y} \right] \frac{\delta_1}{1} \frac{t^2}{2} + \ldots \qquad (7.12b)$$

with $X_s{}^o$, $Y_s{}^o$, $N_s{}^o$ denoting the rudder derivatives of the force-moment system at the steady state. It follows from (7.5) and (7.12) that:

$$Q_R = \frac{Y_\delta{}^o T + N_\delta{}^o T_y}{Y_\delta{}^o T_z T + N_\delta{}^o} \qquad (7.13)$$

This is a dimensional measure of the ship's initial sway sensitivity to a rudder-induced yawing moment.

31

Because the values of $T_y$ and $T_z$ are usually very small, $Q_R$ can be further simplified to the following relation:

$$Q_R = \frac{Y_\delta^\circ T}{N_\delta^\circ} \tag{7.13}$$

## 7.5. Alternative Maneuvering Criteria

An ideal ship preserves its sway speed when maneuvering about a steady straight course. For such a ship, $Q_H'$, $Q_{HR}'$ and $Q_R'$ are each equal to zero. Although it may be unrealistic to identify a ship whose maneuvering indices are zero, it is possible to identify ships whose indices lie within acceptable bounds. Conventional bounds can be derived by using the indices to assess the behavior of ordinary ships. The conventional range of each index is identified by evaluating the indices over a database of 173 ships of international origin representing various design types and sizes dating from the late fifties. The database includes 70 container ships, 66 tankers (6 of which transport oil and natural and petroleum gas), 9 tugs, 7 whalers, 7 ferryboats, 4 naval vessels, 1 freighter, 1 planning-boat, and 1 barge-tug, see Appendix 11. Statistical methods are then used to obtain the desired bounds.

Each ship provides a sample design, and the database is seen as a scattered set of points in a space of possible designs. Performance criteria are developed for each index and are used to partition this scatter diagram into two distinct clusters: a normal group of ships and an abnormal one.

The values of $Q_H'$ and $Q_{HR}'$ that occur normally in practice were identified by solving (7.8) and evaluating (7.11) for each ship in the database. The values of $Q_R'$ were identified by evaluating (7.13). The calculation was performed on a personal computer and employed the hydrodynamic coefficients of Appendix 2 to evaluate (7.9) and (7.10).

The results of the calculation show that the standard deviation of each index is an order of magnitude smaller than the index mean. The nearly uniform distributions of these measures over a wide range of ships and ship types support their use as indicators of normal and abnormal maneuvering behavior.

Bare Hulls Criteria. Conventional bounds for $Q_H'$ at zero trim are calculated from the database. The index lies in the interval $(-.4, -.1)$. Its distribution is approximately normal with a standard deviation of $\sigma = .024$ about the mean $\mu = -.307$, see Figure 3. The average value $\mu = -.307$ most likely reflects an attempt by designers to minimize $Q_H'$ while satisfying other design requirements.

Moreover, the small standard deviation indicates that conventional hulls exhibit acceptable turning behavior only when $Q_H'$ lies near $\mu = -.307$. Thus,

$$\mu \quad \sigma \leq Q_{H'} \leq 0 \text{ or,}$$

$$.331 \leq Q_{H'} \leq 0 \tag{7.14a}$$

provides a unit standard deviation criterion for acceptable bare-hulls at zero trim, and

$$Q_{H'} < .331 \tag{7.14b}$$

is the criterion for unacceptable hulls. Similar bounds can be derived for $Q_{HR}'$ and $Q_R'$.

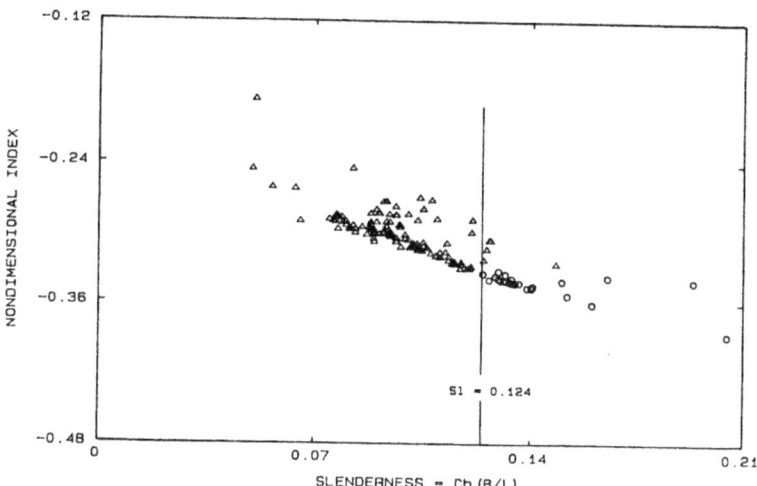

Fig. 3

$Q_H'$ plotted against $S_1$ for each ship in database. Normal hulls that satisfy (7.14a) are indicated by triangles, while abnormal hulls satisfying (7.14b) are indicated by circles.

Bare Hulls with Rudders Criteria. Evaluation of $Q_{HR}'$ over a database of 173 hulls with rudders indicates that this index lies in the range (-.386,-.215).

The distribution of values is approximately normal with a mean $\mu = -.299$ and a standard deviation $\sigma = .032$, see

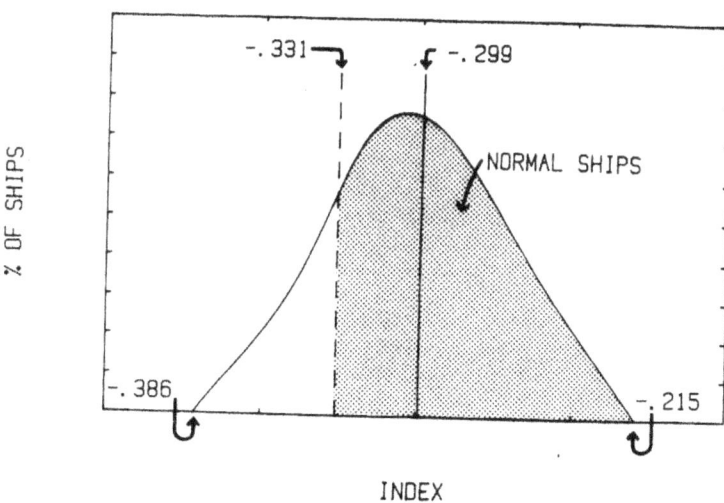

Fig. 4

Statistical distribution of $Q_{HR}'$

Similarly to the bare-hull criteria, the unit standard deviation criterion for conventional zero-trim hulls with rudders is:

$$.331 \leq Q_{HR} \leq 0 \qquad (7.15a)$$

for acceptable turning behavior during transit to a steady straight course, and

$$Q_{HR}' < .331 \qquad (7.15b)$$

33

for unacceptable behavior.

From equations (7.9), (7.10) and (7.11), the considerations on the sensitivity of $Q_{HR}'$ to the variation of L2 given in section 7.3 and the maneuvering criteria outlined in this section, it is possible for $Q_{HR}'$ index to provide the following set of useful information:

| sign | magnitude | manoeuvring performance |
|------|-----------|-------------------------|
| + | < .331 | unstable, acceptable rate of change |
| + | > .331 | unstable, unacceptable rate of change |
| − | < .331 | stable, acceptable rate of change |
| − | > .331 | stable, unacceptable rate of change |

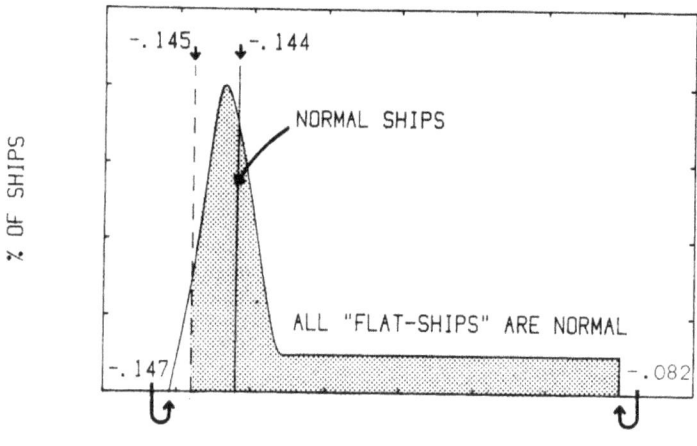

Fig. 5
Statistical distribution of $Q_R'$

Rudder Sensitivity Criteria. The values of $Q_R'$ lie in the interval (-.147, -.082). The index values are distributed about the mean value $\mu$ = -.135 with a standard deviation of $\sigma$ = .011. However, the distribution of conventional values is bimodal, rather than normal, see Figure 5.

91 values are distributed uniformly over (-.142, -.082), and the other 82 values are concentrated in a "spike" supported by the interval (-.147, -.142). The values of $Q_R'$ that lie in the spike define the extreme lower range of conventional values. The spike distribution is nearly normal with a mean of $\mu$ = -.144, and a standard deviation of $\sigma$ = .001. Thus the unit standard deviation criterion for the spike,

$.145 \leq Q_R' \leq 0$ for acceptable hull-rudders   (7.16a)

$Q_R' < .145$ for unacceptable hull-rudders  (7.16b)

provides a conventional rudder performance criterion when exiting from a steady straight course at zero trim. It is noted that (7.16) is based on 82 ships, not the full 173 ships in the database. Because the values of $Q_R'$ of the remaining 91 ships satisfy the proposed criteria, the use of that criteria is justified as a first attempt to define conventional rudder criteria.

### 7.6. An Example for Illustrating the Physical Meaning of the Maneuvering Index

Usually, hulls are designed so that the v'-r' phase diagram is approximately linear near the origin with slope $Q_H'$. An illustration is provided in Figure 6, which is a sketch of the phase diagrams for a submarine with index $Q_H'=-.186$, moving on the surface, and a tug-barge with $Q_H'=-.387$.

Table 2
Principal Dimensions of the Submarine and Tug-barge

|  | L | B | D | cb | Ra |
|---|---|---|---|---|---|
| Submarine | 170.38m | 12.8m | 11.1m | .68 | 33.84m² |
| Tug-barge | 91.44m | 18.9m | 4.72m | .99 | 15.55m² |

The principal dimensions of the vessels are listed in Table 2. The phase diagram is deeper near the origin for the bulky tug than for the slender submarine. The effect of this phase plane behavior on ship trajectories can be seen by comparing Figures 7 and 8.

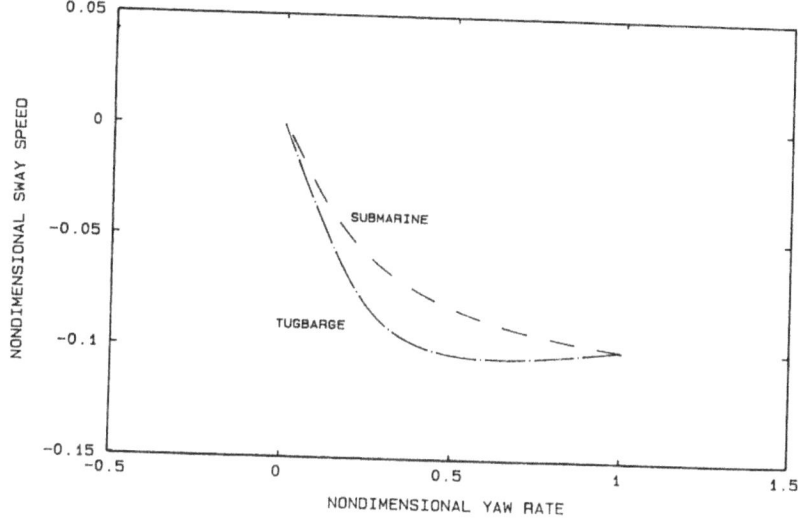

Fig. 6

Phase diagrams for the submarine and tug-barge on approach to a straight course r'=0, v'=0. The vessels initiate their motions at r'=1 and v'=-.1.

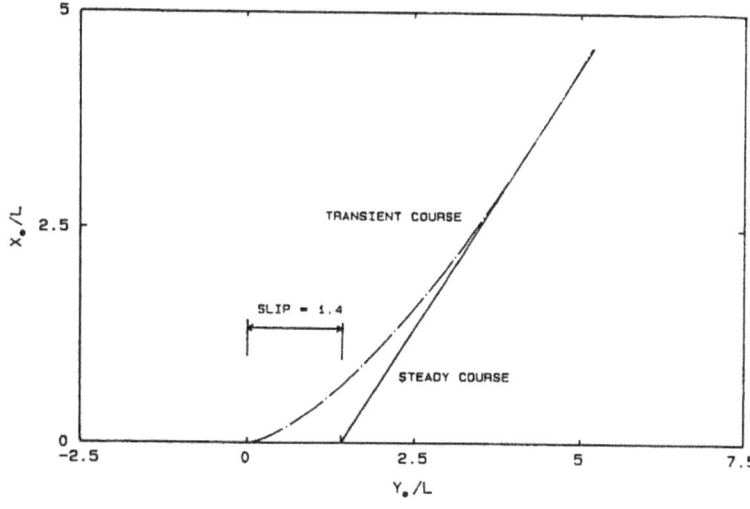

Fig. 7

Tug-barge trajectory on approach to a straight course.
The barge starts at the origin of a space-fixed coordinate system and follows the transient as it nears the steady course.
The vessel slips 1.4 ship lengths during the maneuver.

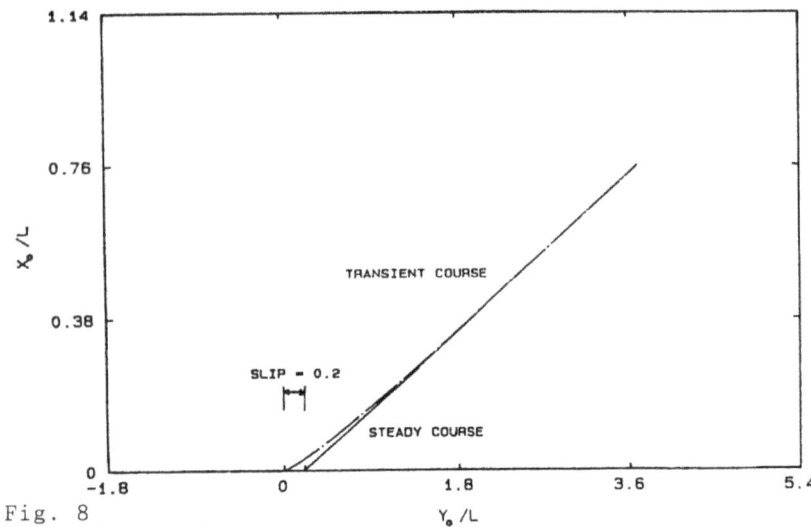

Fig. 8

Fig. 8
Submarine trajectory on approach to a straight course.

The ship starts at the origin of a space-fixed coordinate system and follows the transient as it nears the steady course. The ship slips .2 of its length during the maneuver.

The figures indicate that the tug-barge drifts 1.4 ship lengths before achieving a straight course, while the submarine drifts .2 of its length before assuming a straight course. These figures together with the preceding discussion indicate that maneuverability is enhanced by reducing the magnitude of $Q_H'$.

# 8 PROPOSING AN APPROACH FOR INTEGRATING SAFETY WITH SHIP MANOEUVRABILITY

## 8.1. Basic Requirements

Before proceeding further with the description of the methodology, it is useful to underline clearly the three individual requirements to be satisfied:

- Operational Specifications, with special emphasis on Ship Maneuverability
- An Acceptable Level of Safety
- Cost Effectiveness

Cost-effectiveness should not be confused with low cost. It may be right to pay a higher amount of money for an item or service if it will yield the desired level of effectiveness. For example, a higher maneuverable ship will expect to cost more but it will have lower running cost because of fewer docking problems or of the flexibility to handle a variety of harbors otherwise interdicted. The crucial point is "effectiveness" and the decision on this must be based on both commercial and technical considerations. Despite the fact that cost-effectiveness has been considered for the purpose of the methodology, detailed consideration of cost-effectiveness would be outside the scope of this book.
If the three requirements are combined then to all three will be given equal weighting and the three factors can be incorporated simultaneously and at the right level.

In addition, the starting point of any project must be the identification of the desired goal, which in this study, can be stated as follows:
"To be profitable in the selected operation, by employing a ship with agreed operational specifications, while meeting an acceptable level of safety and maneuverability and at the same time fulfills the demand for cost effectiveness ."

Maneuverability is usually regarded as only one out of several specifications required by the ship owner.
However, maneuverability can greatly affect the range of operations of the ship and can generate hazards to the navigation. In some cases maneuverability is not expressly required as a customer specification. Indeed, the case of a super-tanker which spends most of its time in open sea navigation requires a lesser study of maneuverability at the design stage. On the other hand, for example a ferry or a tug, which a relevant part of the operational time is spent in maneuvering operations, can have higher requirements of maneuvering-safety.
Nevertheless, maneuverability can be important in open sea navigation as a "preventive safety" feature to avoid for example collisions or groundings which, have proved to have catastrophic consequences, see [1, 5 , 6].

The objective is to approach maneuverability in such a fashion that the results would generate, ships which have acceptable maneuvering performances, see Chapter 7 and Appendix 1, related with an acceptable level of safety while meeting the demand for cost-effectiveness.

This requires that maneuvering criteria are incorporated in a proper safety methodology over the appropriate phases of a ship's entire life-cycle. When coupled with maneuvering criteria, there is a safety methodology which appears to satisfy simultaneously the requirements. This methodology, proposed by Kuo, [26, 60-62], is called PREVENT-IT Safety Methodology, see Appendix 7. The basic ideas of the PREVENT-IT Safety Methodology can be readily appreciated if stated as follows:

"To anticipate potential or expected failures by taking active steps to minimize the likelihood of their occurrence or to limit their effects, [26]."

In the ease of this book, the main difference is that the maneuvering aspects will be emphasized. As far it is known, this will be a first attempt to enhance maneuvering-safety using PREVENT-IT.

## 8.2. Proposing an Approach

Ship's Life Cycle.
The first stage is to understand the complete process that begins once the ship owner has outlined the basic requirements. A way of representing it is to examine the "life-cycle" of a ship. The ship's life cycle is regarded as comprising eight phases, see Appendix 6 and Table 3.
Table 3

Table 3

| SHIP'S LIFE CYCLE | |
|---|---|
| Phase | Result |
| Concept Generation | Sound possible concepts |
| Preliminary Design | Meeting the requirements |
| Detailed Design | Specific design for production |
| Procurement | Materials and equipments |
| Construction | Convert design into ship |
| Commissioning | Satisfying specifications |
| Operation | Running the ship |
| Decommissioning | Dismantling the ship |

PREVENT-IT Safety Methodology Integrated with Maneuvering
Criteria. Then, the PREVENT-IT Safety Methodology is employed to evaluate the maneuvering-safety factor in each phase of the life cycle. The PREVENT-IT methodology consists of nine steps, see Appendix 7 and Table 4.
However, it is required that only designs, design variations, ship's samples, and operations which satisfy acceptable maneuverability can be further investigated.
Alternative maneuvering criteria, see Chapter 7, are applied to conform ship samples to acceptable maneuvering performances, in each phase of the ship's life cycle.

Table 4

| PREVENT-IT SAFETY METHODOLOGY | |
|---|---|
| Step | Action |
| 1 | Predict Potential Manoeuvring Hazards |
| 2 | Research into Manoeuvring Risks |
| 3 | Establish the Role of Human Involvement |
| 4 | Verify the Scope for Design Modifications |
| 5 | Engineer the Containment Systems |
| 6 | Nominate Viable Emergency Solutions |
| 7 | Transmit Quality Requirements |
| 8 | Interface with Manoeuvring and Other Regulations |
| 9 | Train the Personnel |

In practice the approach can be applied in many ways.

For the purpose of this study, it is believed that the activities associated with the design phases contribute directly to the inherent maneuverability of the ship. The approach may be confined within the first two designs phases of the ship's life cycle:

• CONCEPT GENERATION PHASE
• PRELIMINARY DESIGN PHASE

However, the choice of the first two phases of the life's cycle doesn't restrict further use of the proposed approach over the entire life cycle which will be the natural path for a total maneuvering-safety program.

# 9 CASE EXAMPLE 1: DRY CARGO SHIP DESIGN

## 9.1. Objective

The overall objective of this case example can be stated as follows:

"The selection and refinement of a dry cargo ship (of Mariner Type) in which there is a requirement for an improved level of maneuverability and safety".

The activities associated with the design phases contribute directly to the inherent maneuverability of the ship, [18]. For this reason, this case example is confined within the first two design phases of the ship's life cycle:

• Concept Generation Phase
• Preliminary Design Phase.

## 9.2. Concept Generation Phase

The aim of this phase can be stated as follows:

"To put forward sound feasible concepts this would meet the customer's requirements"

The output of this phase would be a design concept that satisfies an acceptable level of maneuverability and safety.

The following procedure is used in this phase:

a. To identify a number of feasible concepts generated through brain-storm sessions and from database of past experience and that they all satisfy the maneuvering criteria given in Chapter 7;

b. To apply the first two steps of the PREVENT-IT Safety Methodology to these feasible concepts in order to provide safety data for decision making;

c. To derive the preferred concept based on how well it satisfies the maneuvering-safety criteria.

The other steps of the PREVENT-IT Safety Methodology are not directly relevant in this phase.

a. To identify the feasible concepts. Ideally, the cargo carrying capacity of feasible ship concepts should satisfy the customer's requirements. This would require all concepts to have the same cargo carrying capacity.
However, due to the limitations of the available data for risk analysis purpose, it was not possible to satisfy this requirement and instead the three concepts selected have different dimensions.

Three conventional bare hull concepts of the Mariner Type that satisfy the classification rule $W_H$, see Appendix 1, are identified for further investigation:

| Concept | W$_H$ | L(m) | B(m) | T(m) | cb |
|---------|-------|------|------|------|-----|
| A | -5.15 | 161.00 | 23.20 | 7.50 | 0.58 |
| B | -3.55 | 70.00 | 10.20 | 3.40 | 0.65 |
| C | -4.86 | 100.00 | 14.30 | 4.80 | 0.60 |

b. To apply PREVENT-IT. Each concept will now be examined using the first two steps of PREVENT-IT.

STEP 1: Predict Potential Maneuvering Hazards.
The following list of potential hazards, [16, 74], have been identified:

1. Grounding and stranding
2. Collision with another ship
3. Contacts and impacts, these include: striking or touching a pier, an aid to navigation, a bridge, etc.

STEP 2: Research into Maneuvering Risks.
Risk is defined as the product of probability of occurrence P by consequences probability C, [55]. For this reason, the analysis of maneuvering risks required the determination of P and C. Published statistics, [76], reports that 10.19, 6.64 and 2.08 per 1000 ship-years of dry cargo ships are involved respectively in maneuvering hazard No. 1, 2 and 3. Reports [51, 77] show the distribution of the probability of occurrence P on the ship length of dry cargo ships. For dry cargo ship of length of 161 m such as Concept A, it is shown that 10.6% of the probability of occurrence of maneuvering hazards belongs to this ship length. This means that 1.08 (10.19 x .106), 0.70 and 0.22 per 1000 ship-years are the probabilities of occurrence for the three maneuvering hazards for Concept A. Expressing the probability of occurrence with the following scale:

| Classes of Probability of Occurrence: scale 0 ¦ 1 |
|---|
| 1 (Frequent)    =  3.39 in 1000 ship-years |
| 0 (Improbable)  =  0    in 1000 ship-years |

The previous values of probability of occurrence become 0.31, 0.21 and 0.06 respectively for hazard type No. 1, 2 and 3.

The calculation of the consequences probability C requires first of all defining the severity classes. The value C represents the conditional probability that the consequences with the specified severity class (see Appendix 9, Table 9.3) will occur given that the hazard occurs.
Usually for complex systems such as ships, C is difficult to calculate and thus becomes a matter of judgment, meaning it is greatly driven by the analyst's prior experiences. However, C can be based on the severity of the hazard as shown in Table 9.4 of Appendix 9, ranging from 0 for zero consequences probability to 1 for the total loss or death. Published statistics, [51,75, 77], reports that for each hazard type the total loss percentages are 27.6%, 12.3% and 10.1% respectively for hazard 1, 2 and 3. While the distribution by ship length gives 0% of total loss for Concept A. The derived values of C for Concept A are 0.65, 0.6 and 0.55.

Appendix 9 provides the details of the calculation of the risk and a summary of numerical results is given below.

CONCEPT A

| HAZARD | P | C | R |
|--------|------|------|-------|
| 1 | 0.31 | 0.65 | 0.201 |
| 2 | 0.21 | 0.6 | 0.126 |
| 3 | 0.06 | 0.55 | 0.033 |
| | | Risk Average=0.120 | |

Using a similar technique it is possible to obtain values of P and C for the other two concepts. Appendix 9 provides the details of such calculation.

CONCEPT B

| HAZARD | P | C | R |
|--------|------|------|-------|
| 1 | 0.82 | 0.75 | 0.615 |
| 2 | 0.53 | 0.70 | 0.371 |
| 3 | 0.17 | 0.65 | 0.110 |
| | | Risk Average=0.365 | |

CONCEPT C

| HAZARD | P | C | R |
|--------|------|------|-------|
| 1 | 1.00 | 0.90 | 0.900 |
| 2 | 0.65 | 0.85 | 0.552 |
| 3 | 0.20 | 0.80 | 0.160 |
| | | Risk Average=0.537 | |

c. To derive the preferred concept. From the risk analysis it is possible to decide that Concept A is the preferred one from safety consideration because it has an overall risk average of 0.120 against 0.365 of Concept B and 0.537 of Concept C. Moreover, Concept A has the best value of $W_H$. Concept A is chosen for further design satisfying both maneuvering and safety criteria at the Conceptual stage.

## 9.3. Preliminary Design Phase
The aim of this phase can be stated as follows:

"To establish whether the preferred concept will meet the agreed requirements in the most effective way to enable management decision to be made".

The output of this phase is a preliminary design that will allow the customer to decide whether to go ahead with the proposal or to consider additional fresh concepts to meet the desired objective.

The following procedure is used in this phase:

a. To apply the first four steps of the PREVENT-IT Safety Methodology to the preferred concept in order to provide safety data for decision making;

b. To derive a preliminary design that satisfies the maneuvering-safety criteria.

The other steps of the PREVENT-IT are not directly relevant in this phase.

a. To apply PREVENT-IT. According to the results of the previous phase, the concept to be used for further investigation has the following main characteristics:

L = 161 m; B = 23.2 m; T = 7.5 m;
cb = 0.58; cp = 0.67; cm = 0.87; cw = 0.69

STEP 1: <u>Predict Potential Maneuvering Hazards</u>.
Different techniques can be applied to predict potential maneuvering hazards. Appendix 7 reports that techniques such as HAZOP and HAZAN can be used to identify the hazards. In this case, however, the identification of the potential hazards is based on available data from previous studies, [5, 6, 74]. These studies appraise collisions and contacts to be potential maneuvering hazards. Considering collisions and contacts as one hazard type, [74] provides three different potential hazard's scenarios as follows:

1. Collision and contact in open waters
2. Collision and contact in restricted waters
3. Collision and contact in port

STEP 2: <u>Research into Maneuvering Risks</u>. The potential maneuvering hazards, probability of occurrence and consequences drawn from published data, [1-6, 74], were calculated in Appendix 9 (Section 9.2). The summary of results of the risk analysis are as follows:

CONCEPT A

| HAZARD | P | C | R |
|--------|-------|------|-------|
| 1 | 0.014 | 0.60 | 0.008 |
| 2 | 0.029 | 0.80 | 0.023 |
| 3 | 0.025 | 0.70 | 0.017 |
| | | Risk Average=0.016 | |

Chapter 6 reports that the main causes of most collisions and contacts were due to:

C.1. inherent poor maneuvering capability of the ship;

C.2. faulty maneuvering systems;

C.3. human errors in misjudging or misunderstanding the maneuvering behavior of the ship.

These causes are too general to be used further in the application of the methodology. However, examination of the available data [1-6] permits of expressing the first two causes C.1. and C.2. in other terms as follows:

C.1.a. <u>Hull Reaction</u>: The hull form is such that in some operating conditions the ship does not execute a proper maneuver to yaw stimulation.

C.1.b. <u>Rudder/s Sensitivity</u>: Rudder-propeller-hull Interaction. The combination of type, size and location of rudder/s and propeller/s generate sluggish ship sensitivity to the rudder stimulation. In other words, it means that when the rudder is operated for controlling the ship, the turning force is not promptly generated as expected for that type of ship.

C.2. <u>Maneuvering Control</u>: failure or malfunctionment of one or all rudders, propellers, thrusters and/or the main engine.

STEP 3: <u>Establish Human Involvement Level</u>.
Ship Maneuvering Casualties are the reflection of the loss of control of the work environment in different ways, [1]. If the assumption is accepted that humans tend to resolve their degree of freedom to get rid of choice and decision during normal work and that errors are a necessary part of this adapting, the trick in design of reliable maneuvering ships is to make sure that human operators maintain sufficient flexibility to cope with ship system aberrations, i.e., not to constrain them by an inadequate rule based system, [1, 5, 6, 72]. Moreover, it would be essential for the operator to maintain "contact" with maneuvering hazards in such a way that they will be familiar with the boundary to loss of control and will learn to recover, [72].
In maneuvering, in which the margins between normal operations and loss of control are made as wide as possible, the odds are that the operators
will not be able to sense the boundaries and, frequently, the boundaries will then be more abrupt and irreversible. For example, when radar was introduced to increase safety at sea, the result was not increased safety but more efficient transportation under bad weather conditions.

Another problem is produced by the changing requirements of ship management. Present organization structures and management strategies in maritime industry still reflect a tradition that has evolved through a period when safety could be controlled directly and empirically.

Furthermore, safety depends on management awareness, and management safety systems are essential. Components of such management safety system should include:

a) a complete commitment to safety by every level at management including the directors of the company;

b) the assignment of clear responsibilities;

c) proper access by relevant line management to specialist safety and other technical advices, from staff management and others;

d) the ensuring of individual competence and adequate training;

e) robust and comprehensive safety systems applying to all phases of activities;

f) auditing to ensure that all of this is in place and working.

Indeed, cause C.3. underlines the importance of the human factors in maneuvering hazards. A review of 12 studies found that the percentage of accidents due to human errors ranged from 4% to 90%, [78].

On the other hand, a study of shipping accidents clearly reported that human error was involved in 96% of collisions, [72].

Because humans are involved significantly in accidents, it is essential to provide proper attention to human contributions by the interested parties.

In fact, people contribute in all causes at different level and at different time, [72-78].
For these reasons, it is essential to incorporate human factors along with the technical aspects, when enhancing maneuvering-safety. However, this research has emphasized the technical aspects.

STEP 4: Verify Scope for Design Changes.

According to [18,29], a "poor" inherent maneuverability means that the ship generates an unnecessary additional surge and sway speed.

Indeed, Chapter 7 has underlined that during a maneuver a ship should ideally have the sway speed (v) decaying exponentially to zero, while the yaw rate (r) would decay at the slower rate.

Thus the sway speed is attenuated more rapidly than the yaw rate and the hull turns onto the new course with essentially no drift.

The following possible design changes to minimize the causes are suggested:

```
Design Changes

C.1.a. Hull Reaction:
the hull can be designed to incorporate
proper sway speed variation with respect
to yaw stimulation. Alternatively, lateral
thrusters can be introduced.

C.1.b. Rudder/s Sensitivity:
specific rudder can be designed and
used to generate required hydrodynamic
forces. The rudder-propeller-hull
configuration can be modified to overcome
the problem.  Alternatively specific type
of propeller can be used to improve the
forces.

Manoeuvring Control and Human Errors:
        deserving further study.
```

The implementation of the changes C.1. requires that the three indices discussed in Chapter 7 are minimized in the ship design:

$Q_H'$, $Q_R'$, $Q_{HR}'$.

The selection of the optimum hull form, rudder and propeller needs that the indices of the maneuvering criteria of Chapter 7 are calculated for various design configurations. At the preliminary design phase, the evaluation of the indices requires that the hydrodynamic coefficients are calculated using a calculation method that takes into account the ship geometric relationships and ship form. Appendix 2 presents a calculation technique of the hydrodynamic coefficients developed for the purpose of this research, while Appendix 3 describes a technique developed to approximate hull forms for maneuvering purposes.

**To derive a preliminary design**. Appendix 8 reports the methodology used to implement the design changes. In the Appendix, five different stern designs were tested with simulation and three different rudders together with various rudder-propeller locations. The preliminary design incorporating acceptable maneuverability and an improved level of safety is derived with the main characteristics of Table 5. An illustration of the bare-hull is presented in Figures 9 and 10.

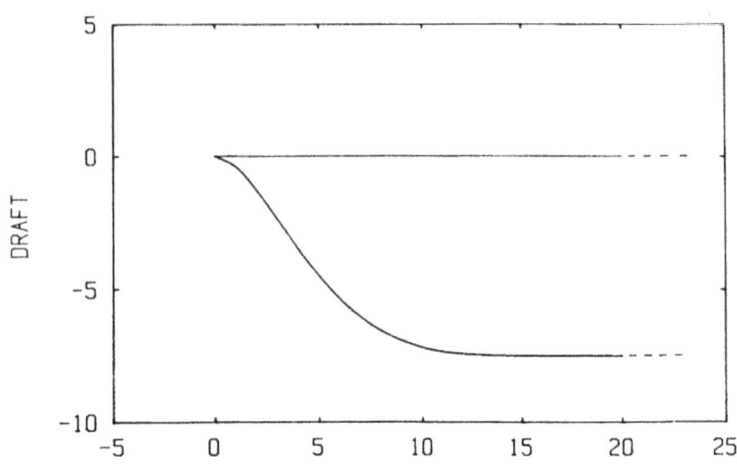

Fig. 9

Adopted bow design.

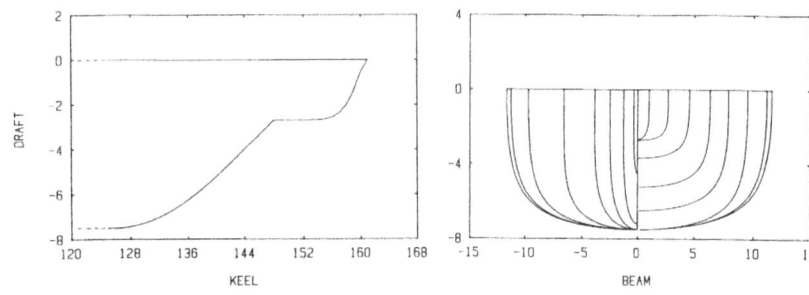

Fig. 10

Adopted Ship design with stern N. 1

Table 5
Main parameters of the derived ship design

$$
\begin{aligned}
\sqrt{Ra}/D &= .535 \\
xr/L &= -.472 \\
xp/L &= -.421 \\
dp/T &= .466 \\
Q_H{}' &= -.290 \\
Q_R{}' &= -.130 \\
Q_{HR}{}' &= -.307
\end{aligned}
$$

## 9.4. Key Results

This example has illustrated how the proposed approach is put into practice in the case of a dry cargo ship design for achieving improved maneuvering-safety features.

The use of the maneuvering classification rules permits earlier concepts to be identified. Later in the PREVENT-IT application, maneuvering criteria are used in the step 4 of the preliminary design to propose scope for design changes. This has been possible, because of the hydrodynamic calculation method developed for the scope of this book, see Appendix 2 together with Appendix 3.

The final output is a preliminary ship design that has the following percentage of improved maneuvering capability when compared with the acceptable level of maneuverability defined in Chapter 7 of this Book:

| MARINER TYPE DRY CARGO SHIP | |
|---|---|
| MANOEUVRING PERFORMANCE | PERCENTAGE OF IMPROVEMENT |
| Bare-hull Index ($Q_H'$) | 12.38% |
| Rudder Index ($Q_R'$) | 10.34% |
| Bare-hull-Rudder-Propeller Index ($Q_{HR}'$) | 7.25% |

# 10 CASE EXAMPLE 2: SERVICE SHIP DESIGN

### 10.1. **Objective**
The overall objective of this case example that is based on a Jackclass Type model can be stated as follows

"To identify and advance the design of a specific service vessel (of Jackclass Type) where the demand for higher maneuverability and safety is an important aspect of its productiveness".

In practice the proposed approach can be applied in many ways. In Chapter 8, it has been underlined that the inherent maneuverability of the ship is defined at the design stage. For this reason, the approach is confined within the first two design phases of the ship's life cycle:

• Concept Generation Phase
• Preliminary Design Phase

### 10.2. **Concept Generation Phase**
The aim of this phase can be stated as follows:

"To put forward sound feasible concepts which would meet the customer's requirements"

The output of this phase would be a design concept that satisfies an acceptable level of maneuverability and safety. Similar to the previous case example, the following procedure is used in this phase:

a. To identify a number of feasible concepts generated through brain storm sessions, based on the available prototype model, and from a database of past experience that they all satisfy the maneuvering criteria given in Chapter 7;

b. To apply the first two steps of the PREVENT-IT Safety Methodology to these feasible concepts in order to provide safety data for decision making;

c. To derive the preferred concept based on how well it satisfies the maneuvering safety criteria.

The other steps of the PREVENT-IT are not directly relevant in this phase.

a. To identify the feasible concepts. In this case example the maneuvering criteria and the Q-indices, see Chapter 7, are still valid. To calculate the Q-indices, it is necessary to use a calculation method of the maneuvering coefficients and a method to approximate hull-concept designs, as shown respectively in Appendices 2 and 3. Three concepts of the Jackclass Type were identified for further study:

| Concept | QH' | L(m) | B(m) | T(m) | cb |
|---------|-------|-------|-------|------|------|
| A | -0.251 | 47.00 | 16.00 | 4.30 | 0.75 |
| B | -0.294 | 40.00 | 14.00 | 3.75 | 0.70 |
| C | -0.249 | 45.00 | 15.00 | 4.00 | 0.72 |

All three concepts have the bare hull Q-index within the range of acceptable maneuvering performances as defined in Chapter 7.

b. To apply PREVENT-IT. Each concept will now be examined using the first two steps of PREVENT-IT.

STEP 1. Predict Potential Maneuvering Hazards.
The following potential maneuvering hazards were identified, [2, 4]:

1. Collision with another ship
2. Contacts with a mobile installation
3. Contacts with a fixed installation

STEP 2. Research into Maneuvering Risks.
Using a procedure similar to the previous case example, see Appendix 9, the risk is calculated on the basis of published data in [2] and [4], that contains also distributions of casualties on the ship size. The published data are then transformed in the severity category and consequences probability of Appendix 9. While, the probability of occurrence scale is:

```
Classes of Probability of Occurrence: scale 0 ¦ 1

    1 (Frequent)    =  1.61  in 1000  ship-years
    0 (Improbable)  =  0     in 1000  ship-years
```

The summary of the results of the risk analysis performed from [2, 4] are as follows:

CONCEPT A

| HAZARD | P | C | R |
|--------|-----|-----|------|
| 1 | 0.4 | 0.4 | 0.16 |
| 2 | 0.6 | 0.6 | 0.36 |
| 3 | 0.3 | 0.4 | 0.12 |
| | | Risk Average=0.213 | |

CONCEPT B

| HAZARD | P | C | R |
|--------|-----|-----|------|
| 1 | 0.8 | 0.3 | 0.24 |
| 2 | 0.5 | 0.6 | 0.30 |
| 3 | 0.3 | 0.5 | 0.15 |
| | | Risk Average=0.230 | |

CONCEPT C

| HAZARD | P | C | R |
|--------|-----|-----|------|
| 1 | 0.5 | 0.6 | 0.30 |
| 2 | 0.7 | 0.4 | 0.28 |
| 3 | 0.3 | 0.4 | 0.12 |
| | | Risk Average=0.233 | |

c. To derive the preferred concept. From the risk analysis it is possible to deduce that Concept A is the preferred one for safety considerations because it has an overall risk average of 0.213 against 0.230 of Concept B and 0.233 of Concept C. Moreover, it has been found that the risk of hazard No. 2 is the most relevant to be minimized because of a value of 0.36. Moreover, Concept A has also good maneuvering index $Q_H'$. Concept A is chosen for further design satisfying maneuvering and safety criteria at the Conceptual stage.

## 10.3. Preliminary Design Phase

The aim of this phase can be stated as follows:

"To establish whether the preferred concept will meet the agreed requirements in the most effective way to enable management decision to be made".

The output of this phase is a preliminary design that will allow the customer to decide whether to go ahead with the proposal or to consider fresh concepts to meet the customer's requirements.

The procedure used in this phase is as follows:

a. To apply the first four steps of PREVENT-IT Safety Methodology to the preferred concept in order to provide data for decision making;

b. To derive the preliminary design including the design changes of the previous step.

It is believed that the other steps of the PREVENT-IT Safety Methodology are not directly relevant in this phase.

a. To apply PREVENT-IT. According to the previous phase and an available Jackclass prototype model, the following concept is now examined:

L=47.00 m; B=16.00 m; T=4.30 m;
cb=0.75; cp=0.767; cm=0.923; cw=0.825.

STEP 1. Predict Potential Maneuvering Hazards.
From a detailed analysis of previous data of service ships, [4], the potential hazards of Concept A are identified and

grouped in the following 4 scenarios:

1. heave collision or contact at the stern;

2. collision or contact when alongside (moored or joystick positioned):
a) stern surge collision or contact
b) stern sway collision or contact
c) side sway collision or contact

3. collision or contact when approaching or departing

4. collision or contact of drifting vessel:
a) sideways drifting: impact amidships
b) sideways drifting: bow or stern impact
c) forward drifting: bow impact

STEP 2. Research into Maneuvering Risks.
Information on past collisions of service ships (attendant vessels) with offshore structures including 107 reports of collisions and contacts are available from [4]. In this case, the risk analysis is faced with a collection of observed statistics. The observations in [4] may not follow any known and well-established probability distributions.

However, these data can be represented through empirical distributions, [53].

The observed collisions for each scenario are reported in [4]. The measure of interest is the probability of collision associated with each scenario of a service ship. This is obtained dividing each scenario frequency by the total number of ships (107). Consequences for each scenario are published in [4]. Then, the probability of occurrence and consequences probability are calculated with a procedure similar to Appendix 9.

The summaries of results of the risk analysis performed from the data in [4] are as follows:

CONCEPT A

| HAZARD | P | C | R |
|--------|-----|-----|------|
| 1 | 0.1 | 0.7 | 0.07 |
| 2 | 0.7 | 0.1 | 0.07 |
| 3 | 0.5 | 0.9 | 0.45 |
| 4 | 0.2 | 0.5 | 0.10 |
| | | Risk Average=0.172 | |

From the risk analysis it is possible to deduce that hazard 3 is the most relevant and it needs to be minimized. Its risk value of 0.45 is about four times higher than any other three hazards.
The causes, see [4], were identified as follows:

C.1. Active Control Causes: failure of functioning or malfunctionment of one or all rudders, propellers, thrusters and/or the main engine.

C.2. Passive Control Causes: the design of the hull is such that in some operating conditions the ship has a sluggish or too rapid ability in changing or preserving direction and speed.

C.3. Active-Passive Control Causes: wrong positioning, dimension, proportion, number, and type of rudders, propellers, thrusters, and/or position of ship's center of gravity may generate failures in maneuverability.

C.4. Hull-Environment Interaction Causes:
presence of waves, wind, current, shallow water, bank, body, may reduce or change the maneuverability of a ship.

Collision or contact impact force is function of the ship motion. On the other hand, from the available data, [4], it is readable that contacts while maneuvering are the most severe because of the approaching speed.

STEP 3. Establish Human Involvement Level.
The presence of human operators operating the service ship makes this a high priority. The preliminary design phase can contribute to minimize the level of human involvement. The considerations of STEP 3 of Case Example 1 still apply here. Indeed, people contribute in all causes at different level and at different time, [72-78], except C.4. For example, relevant aspects of human factors include at least the following:

• Ship Designers: Responsibility of Design Solutions which influence the maneuvering performances;

• Ship Operators: Responsibility in controlling the maneuvering behavior of the ship;

STEP 4. Verify Scope for Design Changes.
The objective here is to verify scope for design changes so that proper ship maneuverability can minimize the frequency of collisions and contacts. The following design changes were proposed to minimize the causes:

```
Design Changes

C.1. Active Controls:  use of a Joystick
control to coordinate the control of rudder/s
and propeller/s. In addition, the following design
improvement can be done:

          1. position and type of rudders
          2. position and type of propellers
          3. position and type of thrusters

C.2.2. Passive control: inherent hull
manoeuvrability requires a hull design which
take into account enhanced manoeuvring performances

C.2.3. Active-Passive Controls:

          1. Avoid the use of side thrusters
          2. Use Propeller at constant RPM
          3. Improve location of turning force
          4. Increase Rudder Size
          5. Use special rudders
```

**To derive a preliminary design**. The objective of this section is to describe how the design changes are implemented in the prototype model. The design changes proposed are aimed to generate a ship that has the following maneuvering characteristics:

a) Change of course rapidly with minimum use of sea room;

b) Maneuver at slow speed, stationary and starting from the rest;

c) Retain control whilst accelerating/decelerating, also in wave-wind condition;

d) Minimize the risk of breakage of the propulsion system.

These characteristics can be achieved as follows:

• Using propellers at constant RPM;

• Fitting special rudders that can generate a hydrodynamic force up to stop the vessel and to satisfy the maneuvering criteria.

In fact, the prototype hull form is fitted with a double vectwin rudder system, see Appendix 5. Figures 11 and 12 illustrate the derived design.

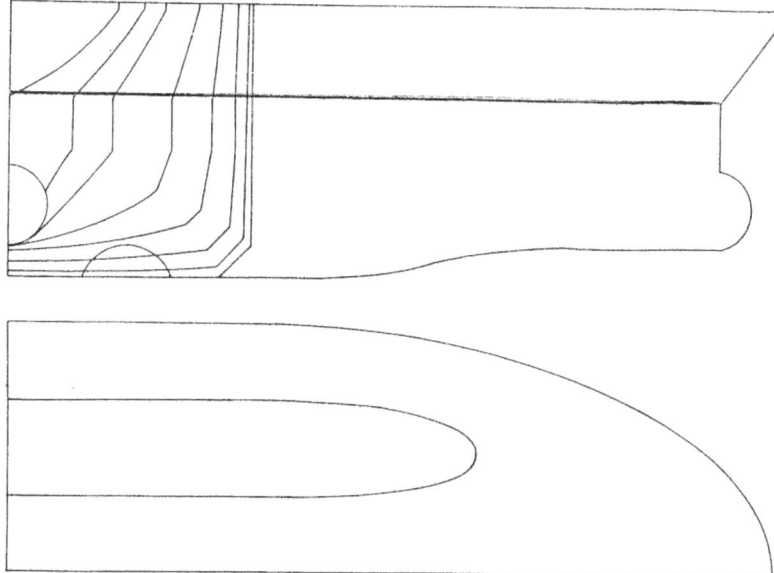

Fig. 11
Body Plan of the design. Derived bow.

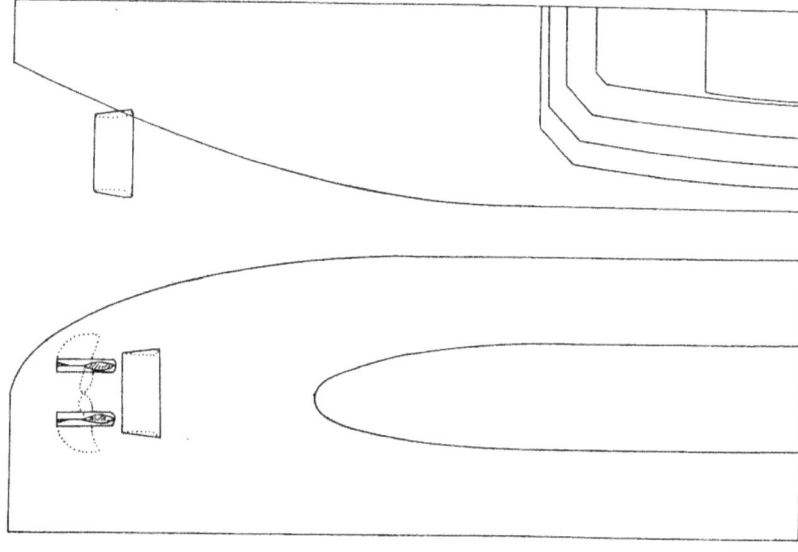

Fig. 12

Body Plan of the design. Derived stern.

10.4. **Key Results**

The ship design generated by the two early Design Phases has shown the application of the proposed approach for a design that uses a non-conventional hull and an innovative maneuvering system. Figures 13 and 14 illustrate the comparison between the bare-hull maneuvering index and the rudder index of the derived design and the indices of the 173 ships in the database.

The prototype model was tested with free running trials at Strathclyde University, see Appendix 5. The maneuvering system of double vectwin rudder system and two constant speed one-directional fixed propellers provides vessel control at all times, going ahead, astern, and sideways, without bow thrusters, gearboxes or other propellers. Appendix 5 provides a description of the double vectwin rudder system and a comparison between the results obtained with simulation and the free running model measurements. As an example, Figure 15 shows the comparison between a computer simulation trial and an experimental measurement.

The derived design has the following maneuvering indexes:

$Q_H' = -.251$

$Q_R' = -.073$

$Q_{HR}' = -.306$

If a skeg is employed, the value of $Q_{HR}'$ increases to

$Q_{HR}' = -.372$

(skeg location at xst=-18.45 m from the center of gravity, and with an height hs=1.98 m and chord cs=7.56 m, and skeg area Ass=7.48 m²). Often, the use of a skeg is related to the improvement of the straight course stability. However, in this particular example the presence of 4 rudders, of a total area = 14.616 m² , about double than the skeg area, makes the skeg unnecessary.

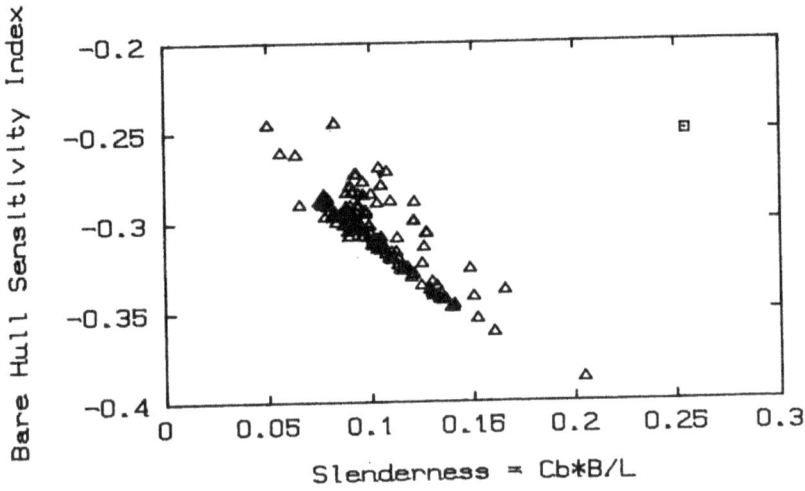

Fig. 1 3

A scatter diagram in the design space spanned by the slenderness (cb*B/L) and the Bare Hull Sensitivity Index ($Q_H'$). Each triangle represents a ship in a database of 173 ships. The square represents the design under consideration.

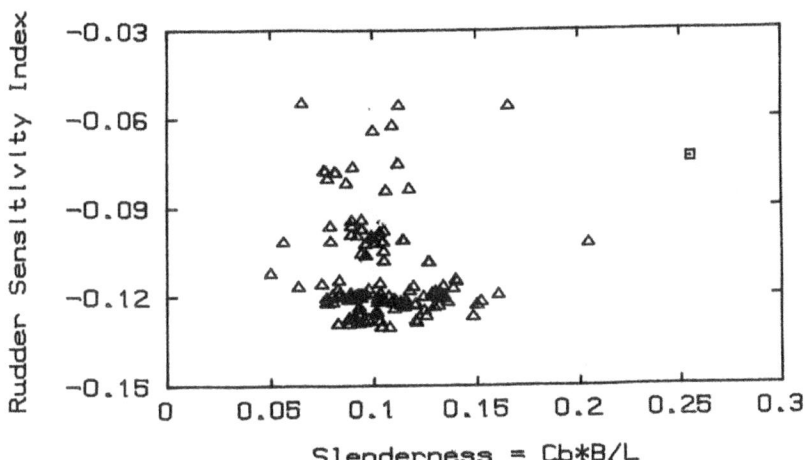

Fig. 14

A scatter diagram in the design space spanned by the slenderness (cb*B/L) and the Rudder Maneuvering Index ($Q_R'$). Each triangle represents a ship in a database of 173 ships. The square represents the design under consideration.

Photo 1

Photo 2

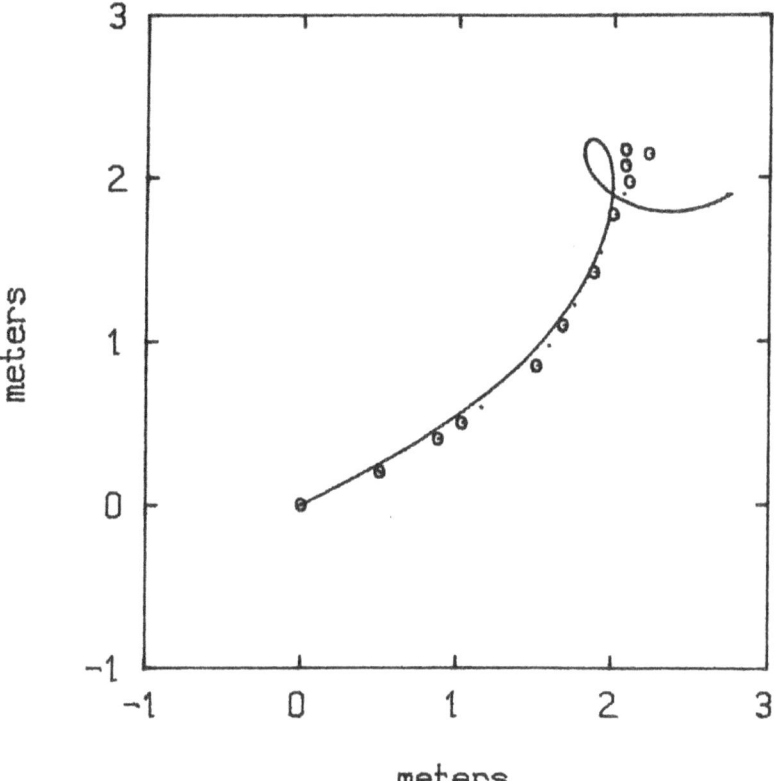

Fig. 15
Model trajectory on a turning circle.
Free running model test = circles;
simulation = solid line. Approaching speed = 1.59 m/s;
joystick angle= -45.

Figure 11 illustrates that the service ship is more "fat" than any other of the 173 ships, but the absolute value of the index (square on the figure) is very close to zero which means that the proposed design sensitivity to react to a rudder stimulation is better than most of the 173 ships in the database. This ship design has the following percentage of improved maneuvering capability when compared with the acceptable level of maneuverability defined in this book:

| JACKCLASS TYPE SERVICE SHIP | |
|---|---|
| MANOEUVRING PERFORMANCE | PERCENTAGE OF IMPROVEMENT |
| Bare-hull Index ($Q_H'$) | 24.16% |
| Rudder Index ($Q_R'$) | 49.65% |
| Bare-hull-Rudder-Propeller Index ($Q_{HR}'$) | 7.5% |

# 11. DISCUSSION

On the basis of the methodology presented in this book, it would be useful to examine the following issues:

<u>Importance of Decisions Made at the Design stage</u>. A vessel's maneuverability is established at the design stage either directly or indirectly. Even if maneuverability is not explicitly considered by the designer, the selected hull form, principal dimensions and characteristics, rudders and propellers will dictate the maneuvering capabilities. Indeed, inherent maneuverability is defined by choices made at the design stages. Particularly, decisions taken during the concept generation and preliminary design phases are of primary importance to the final ship maneuvering capability.

Although modifications can be and sometimes are made during the detailed design and construction phases, these are usually only second order changes. Of course, maneuverability is only one of the many design requirements which must be met. However, it is possible to expect that maximum improvement of ship maneuverability would be commensurate with the other design requirements.

<u>Role of Human Factors</u>. Many studies have been proposed to improve maneuvering-safety, e.g. newer design ideas, better technical aids, more stringent regulations.

However, the present rate of maneuvering casualties is still unreasonably high. Available analyses of accidents have classified the cause of an accident in two main groups:

- human errors;
- faulty systems.

The relative importance of these two factors varies greatly with work situations. For example, a review of 12 studies found that the percentage of accidents due to human errors ranged from 4% to 90%, [78]. On the other hand, a study of shipping accidents clearly reported that human error was involved in 96% of collisions, [72].

Whatever is the right percentage of human errors, one of the main causes of maneuvering casualties still remains the human factors. Because humans are involved significantly in accidents, it is essential to provide proper attention to human contributions by the interested parties, they include:

- Ship Designers : Responsibility for Design solution
- Ship Operators: Controlling the ship Maneuvers
- Ship Owners: Management Policies
- Regulators: Devising Regulations

For these reasons, it is essential to incorporate human factors along with the technical aspects, when enhancing maneuvering-safety. Since a human factor is a broad subject, this book has emphasized on the technical aspects with only limited attention being paid to the role of human factors.

<u>Book Contributions</u>. The present book has made the following main contributions:

(a) Providing fresh understandings of the term "Ship Maneuverability" and what are acceptable Maneuvering qualities, and demonstrating how Safety can be incorporated with Ship Maneuverability;

(b) Proposing a fresh approach that would allow Safety to be integrated with Ship Maneuverability at the design stage of a ship's life cycle;

(c) Illustrating with two case examples (a Mariner Type dry cargo ship and a Jackclass service ship) how the proposed approach can be put into practice at the conceptual and preliminary design phases and how improved maneuvering-safety features can be achieved;

<u>Areas Requiring Further Attention</u>. When applying the proposed approach, additional care is needed in areas of practical interest, including the following:

(a) To examine how best to incorporate human factors into ship maneuvering-safety during the application of the proposed approach;

(b) To enhance the quality of the information stored on the database used in devising the maneuvering criteria; it is desirable to acquire more ship and model scale maneuvering data and relevant design parameters;

(c) To seek ways that will enable the effects of trim and environmental conditions to be introduced into the proposed criteria and into the maneuvering performance prediction method;

(d) To illustrate how the methodology of this book at the design stage affects the application of the proposed approach at the operational phase;

# 12 CONCLUSIONS

Based on the contents of this book the following conclusion can be drawn:

a) Critical reviews performed on available approaches for treating ship maneuverability, safety and methods of incorporating safety features with maneuverability indicated that there is a wide scope for devising improved techniques.

b) A fresh approach for integrating Safety with Ship Maneuverability has been proposed and it involves examining maneuvering and safety features in the appropriate phases of a ship's life-cycle while taking into account aspects of engineering, management and operation.

c) Practical applications of the proposed approach have demonstrated that it is at the concept generation and the preliminary design phases that maximum improvement can be achieved.

d) In the two case examples, significant improvements were achieved in the acceptable level of maneuverability as determined in this research project, through the use of the proposed approach.

While satisfying safety requirements, the actual improvement for the Jackclass service ship is about 27% and for the Mariner dry cargo ship is about 10%.

# 13 REFERENCES

1. Penow C., Normal Accidents, Basic Books, New York, 1984

2. Hollingdale S.H. (ed.), Mathematical Aspects of Marine Traffic, Academic Press: London, 1979

3. Cahill R.A., Collisions and Their causes, Fairplay Publications, 1983

4. National Maritime Institute, Collisions of Attendant Vessels with Offshore Installations, Part 1 and Part 2, Offshore Technology Report OTH 84208, Department Of Energy, London, 1985

5. Marine Accident Investigation Branch Department Of Transportation, Report 1989, London

6. Johnson R.E. and Katcharian L.Z., "Several Recent Rammings Investigated by the National Transportation Safety Board, Marine Technology, Vol. 28, No.6, Nov. 1991.

7. Kuo C., Business Fundamentals for Engineers, McGrawHill Book Company, 1992

8. E.V. Lewis, Principles of Naval Architecture, Vol. 3, Chapter IX, Ed. Society of Naval Architects and Marine Engineers, Jersey City, N.J., USA, 1989

9. Biancardi C.G., "Practical Calculation Method of Ship Maneuvering Characteristics at the Design Stage," International Shipbuilding Progress, Vol. 37, pp. 221-245, 1991

10. Burcher R.K., "Developments in Ship Maneuverability," The Naval Architect, The Royal Institution of Naval Architects, London, U.K., 114, 1, 1972

11. Kijima K., "Prediction Method for Ship Maneuvering Performance in Deep and Shallow Waters," Workshop on Modular Maneuvering Models, The Society of Naval Architects and Marine Engineers, November 13, 1991

12. Barr R.A., "Estimation of Ship Maneuverability Using Ship Trials Data Bases," Proceedings of the International Conference on Ship Maneuverability, Paper No.7, The Royal Institution of Naval Architects, London, 1987

13. Clarke D., "Assessment of Maneuvering Performance," Proceedings of the International Conference on Ship Maneuverability, Paper No.3, The Royal Institution of Naval Architects, London, 1987

14. Fujino M, Kijima K. and Hamamoto M., "Present State of the Prediction Method of Ship Maneuverability," Proceedings of MARSIM and ICSM 90, pp. 7-17, The Society of Naval Architects of Japan, Tokyo, Japan, 1990

15. The Research Committee of Dynamic Performance Maneuvering and Contro 1 Section, "Prediction of Maneuverability of a Ship," Bulletin of The Society of Naval Architect of Japan, No. 668, February 1985

16. Mikelis N.E. and Prince W.G., "Calculations of Acceleration Coefficients and Correction Factors Associated with Ships Maneuvering in Restricted Waters: Comparisons between Theory and Experiments, " Published for Written Discussion, The Royal Institution of Naval Architects, London, U.K., 1980

17. Coates G.A., "A Shiphandler's View," Proceedings of the International Conference on Ship Maneuverability, Paper No.1, The Royal Institution of Naval Architects, London, 1987

18. Landsburg A.C. et alii, "Design and Verification for Adequate Ship Maneuverability," Transactions of the Society of Naval Architects and Marine Engineers, Vol.91, USA, 1983

19. Doerffer J.W., "Activities of the International Maritime Organization (IMO) in the Field of Manoeuvrability of Ships," Proceedings of MARSIM and ICSM 90, pp. 3-4, The Society of Naval Architects of Japan, Tokyo, Japan, 1990

20. Eda H., Falls R., Walden D.A., "Ship Maneuvering Safety Studies," Transactions of SNAME, Vol. 87, pp. 229-250, 1979

21. Roseman D.P. (editor), The Marad Systematic Series of Full Form Ship Models, The Society of Naval Architects and Marine Engineers, New Jersey, USA, 1987

22. Ankudinov V.K., Controllability Assessment Using Computer Techniques, Maneuvering Design Workbook, Chapter 6, Project Supported by SNAME Panel H-10 (Ship Controllability), Private Correspondence, 1991

23. Kempf G., "Maneuvering Standards of Ships," Deutsche Shiffarts Geirschrifft "Hansa", No. 27/28, 1944

24. Spyrou K. and Vassalos D., "Recent Advances in the Development of Ship Maneuvering Standards," Proceedings of MARSIM and ICSM 90, pp. 51-58, The Society of Naval Architects of Japan, Tokyo, Japan, 1990

25. U.S. Department of Transportation, IMO Ship Maneuverability Standards, Committee Correspondence, SNAME Panel H-10, April 9, 1992

26. Kuo C. ,"A Preventive Framework far Achieving Effective Safety," Proceedings of STAB 90 (Stability of Ships and Ocean Vehicles), Naples, Italy, 1990

27. Barr R.A. and McCoy H.H., "Assessment of Numerical Measures Proposed far Use in Ship Maneuvering Performance Standards Using Data in the U.S. Coast Guard Ship Maneuvering Data Base," Final Report, Panel H-10 (Ship Controllability) of the U.S. Society of Naval Architects and Marine Engineers, April 1991

28. Daidola J.C. and G. Daniel, "Maneuvering in The Ship Design Spiral," Paper presented to: New York Metropolitan Section of The Society of Naval Architects and Marine Engineers, New York, USA, March 11, 1981

29. Biancardi C.G. and Dellwo D.R., "On the Maneuvering Qualities of Ships," Proceedings of the Ninth Ship Control Systems Symposium, Vol. 2, pp. 2.81-2.89, Bethesda, Maryland, USA, 1990

30. Saunders H., Hydrodynamics in Ship Design, Vol. 3, The Society of Naval Architects and Marine Engineers, USA, 1965

31. Lamb H., Hydrodynamics (6th edition), Cambridge University Press, 1932

32. Newman J.N., Marine Hydrodynamics, Cambridge (Mass) MIT Press, 1977

33. Flagg C.N. and Newman J.N., "Sway Added Mass of Rectangular Profiles in Shallow Water," Journal of Ship Research, No. 15, pp. 257-265, 1971

34. Mikelis N.E. and Prince W.G., "Two Dimensional Sway Added Mass Coefficients for Vessels Maneuvering in Restricted Waters," Transactions of The Royal Institution of Naval Architects, No. 121, pp. 145-150, 1979

35. Mikelis N.E. and Price W.G., "Calculation of Hydrodynamic Coefficients for a Body Maneuvering in Restricted Waters using a Three Dimensional Method," Originally Published for Written Discussion, The Royal Institution of Naval Architects, 1980

36. Newman J.N., "Theoretical Methods in Ship Maneuvering," Proceedings of Advances in Marine Technology, Trondheim, Norway, 1979

37. Bishop R.E.D., Burcher R.K. and Price W.G., "The Determination of Ship Maneuvering Characteristics from Model Tests," Published for Written Discussion, The Royal Institution of Naval Architects, London, U.K., 1974

38. F. Shangyong, L. Shaokang and C. Zhiping, "Oblique Towing Test Results of a Series of Twin-screw, Twinrudder Ship Models," Shipbuilding of China, Transactions of the Chinese Society of Naval Architecture and Marine Engineering, No. 100, January 1988

39. Eminente C. and Coppola, "Computational Evaluation of Ship Maneuvering Performance Based on Scale Model Tests," INSEAN Technical Report No. 1985-42, The Towing Tanks of Rome, Italy, 1985

40. Gadd G.E., "A Calculation Method for Forces on Ships at Small Angles of Yaw," Published for Written Discussion, The Royal Institution of Naval Architects, London, U.K., 1986

41. Inoue S., Hirano M., Kijima K., "Hydrodynamic Derivatives on Ship Maneuvering," International Shipbuilding Progress," Vol. 28, No. 321, May 1981

42. Fujina M. and Ishigura T., "A Study of the Mathematical Model Describing Maneuvering Motions In Shallow Water -Shallow Waters Effects on Rudder Effectiveness Parameters, in Japanese," Journal of The Society of Naval Architects of Japan, Vol. 156, pp. 180-192, 1984

43. Dand I.W., "On Modular Maneuvering Models," Proceedings of International Conference of Ship Maneuverability, The Royal Institution of Naval Architects, London, U.K., 1987

44. Oltmann P and Sharma S.D., "Simulation of Combined Engine and Rudder Manoeuvres Using an Improved Model of Hull-Propeller-Rudder Interactions," Proceedings of 15th ONR Symposium on Naval Hydrodynamics, 1984

46. ASTEO, Risk Analysis far Offshore Structures and Equipment, Published by Graham and Trotman, 1987 45. Trankle T.L., "Development of Kings Pointer Maneuvering Model from Sea Trials Data," Final Report far U.S. Department of Transportation, Maritime Administration, Contract No. MA-80-SAC-01092, Washington, USA, August 1991

47. Charlton R.M., "Safety in Exploration and Production Operations," Proceedings APEA, Australia, 1989

48. Lees F.P., Loss Prevention in the Process Industries, Vol. l, Chap. 8, Butterwhorts, London, 1980

49. Neilsen D.S., "The Use of Cause-Consequence Chart in Practical System Analysis, Reliability and Fault Tree Analysis," SIAM, 1972

50. Francescutto A., "Is It Really Impossible to Design Safe Ships?" Spring Meetings of The Royal Institution of

Naval Architects, London, 1992

51. Wahl J .E., "Investigation into the Survival Capability of RO/RO Vessels," Det Norske Veritas Paper Series, Paper No. 839007, April 1983

52. Hutehison B.L., "Risk and Operability Analysis in the Marine Environment," Transactions of SNAME, Vol. 89, pp. 127-154, 1981

53. Modarres M., What Every Engineer Should Know about Reliability and Risk Analysis, Marcel Dekker Ine., New York, 1993

54. Gloss D.S. and Wardle M.G., Introduction to Safety Engineering, John Wiley & Sons: New York, 1984

55. Roland H.E. and Moriarty B., System Safety Engineering and Management, John Wiley & Sons: New York, 1990

56. International Maritime Organization (IMO), Resolutions A.209 and A.160

57. International Maritime Organization (IMO), "Recommendation on the Prevision and the Display of Maneuvering Information on Board Ships," IMO Document A 15/Resolution 601, January 1988

58. International Maritime Organization (IMO), "Interim Guidelines for Estimating Maneuvering Performance in Ship Design,", IMO Document MSC/Circ. 389, January 1985

59. Lloyd's Register of Shipping, "Provisional Rules for the Classification of Ships Maneuvering Capability," October 1986

60. Kuo C., "Recent Advances in Marine Design and Application," ICCAS 91, North Holland Publications, 1991

61. Kuo C., "A Methodology for Evolving Costs and Safety Effective Ships," First Joint Conference on Marine Safety and Environment, Technical University of Delft, Delft, The Netherlands, June 1992

62. Kuo C., "Applying the Prevent-It Safety Methodology to Ship Design Activities," PRADS 92, Eisevier Publications, 1992

63. Biancardi C.G. and Dellwo D.R., "Manoeuvering Criteria at the Design Stage," Practical Design of Ships and Mobile Units, Eisevier Publications, 1992

64. Biancardi C.G. and Dellwo D.R., "Classification of Ships by Their Maneuvering Characteristics," Transactions of The Society of Naval Architects and Marine Engineers, USA, 1991

65. Nomoto K., Taguchi T., Honda K. and Hirano S., "On the Steering Qualities of Ships," International Shipbuilding Progress, Vol. 4, 1957, pp. 354-370.

66. Fujii H., "Proposal on a Practical Prediction Method of Ship Maneuvering Motions by Use of the Index of Manoeuvrability," Proceedings of MARSIM and ICSM 90, Tokyo, Japan, Society of Naval Architects of Japan, 1990.

67. Norrbin, "Ship Handling Standards - Capabilities and Requirements," Proceedings of MARSIM and ICSM 90, Tokyo, Japan, Society of Naval Architects of Japan, 1990.

68. Koyama T. and Kose K., "Recent Studies and Propasals pf the Manoeuvrability Standards," Proceedings of MARSIM and ICSM 90, Tokyo, Japan, Society of Naval Architects of Japan, 1990.

69. Hamamoto M., Honda K. and Iida T., "On the Directional Stability of a Ship During Stopping Maneuver," Proceedings of MARSIM and ICSM 90, Tokyo, Japan, Society of Naval Architects of Japan, 1990.

70. Fukuto J. and Fuwa T., "A Proposal of a Practical Ship Manoeuverability Test," Proceedings of MARSIM and ICSM 90, Tokyo, Japan, Society of Naval Architects of Japan, 1990.

71. Lloyd E. and Tye W., Systematic Safety, Civil Aviation Authority, Landon, 1982

72. Marine Directorate, Department of Transport, The Human Element in Shipping Casualties, 1991

73. Nobukawa T., Kato T., Motomura K. and Yoshimura Y., "Studies an Manoeuvrability Standards From the Viewpoint of Marine Pilots," Proceedings of MARSIM and ICSM 90, pp. 59-66, Tokyo, Japan, Society of Naval Architects of Japan, 1990

74. Young W., "Determining Trends from VTS Observations," Proceedings of Seventh International Symposium an Vessel Traffic Services, Vancouver, British Columbia, Canada, June 8-12, 1992.

75. Mackey T.P, Bingham V.P., Hederstrom A., "Application Experiences with High Maneuverability Rudders," Proceedings of the Ninth Ship Control Systems Symposium, Vol. 1, pp. 1.132-1.168, Bethseda, Maryland, D.S.A., 1990

76. Aldwinckle D.S. and McLean D., "A Safety Review of Ships far Liquified Gases and Future Legislative Needs," Proceedings of GASTECH 84, Amsterdam, November 6-9, 1984

77. Jansson B.O., "Safety of RO-RO Vessels - RO-RO Vessels' Casualty Statistics," The 5th International Conference and Exhibition an Marine Transport using Roll-on/Roll-off Methods, Det Norske Veritas, 1981

78. Sanders M.S. and McCarmick E.J., Human Factors in Engineering and Design, McGraw-Hill International: New York, 1987

79. James M., Classification Algorithms, John Wiley and Sons, New York, U.S.A., 1985

80. Strumpf A., "Prediction of Coefficients and Motions of Submarines at Deep Submergence (U)," Davidson Laboratory Report 1891, Stevens Institute of Technology, Hoboken, NI, U.S.A., September 1976

81. Eskigian N.M. and Sedlak C., "Model Studies of a Systematic Series of StreamIine Bodies of revolution," Davidson Laboratory Report 547, Stevens Institute of Technology, Hoboken, NI, U.S.A., April 1955

82. Van Dyck R.L. and Dugoff H., "Model Tests of a Series of Submarine-Tanker Forms," Davidson Laboratory Letter Reports 781 and 783, Stevens Institute of Technology, Hoboken, NI, U.S.A., February 1960

83. KeIIy H.R, "The estimation of Normal Force, Drag and Pitching Moments for Blunt-Based Bodies of Revolution at Large Angles of Attack," Journal of the Aeronautical Sciences, Val. 21, No.8, August 1954.

84. Lanwerber L., "Added Mass of Lewis Forms Oscillating in a Free Surface," Proceedings of the Symposium the Behavior of Ships in a Seaway, The Netherlands, 1957

85. Lanwerber L. and Macagno, "Added Masses of Two-Dimensional Forms by Conformal Mapping," Journal of Ship Research, SNAME, O.S.A., June 1967

86. Kuo C., Computer Methods for Ship Surface Designs, Longman: London, 1971

87. Private Correspondence with SNAME Panel H-10, Some Results for a Comparison of ESSO OSAKA Simulation Model, The Marine Board Committee on Assessment of Shiphandling Simulation, O.S.A., September 1991

88. Eda H. and Crane C.L., "Steering Characteristics of Ships in Calm Water and Waves," Transactions of SNAME, Vol. 73, 1965

89. Holtrop J. and Mennen G.G.J., "An Approximate Power Prediction Method," International Shipbuilding Progress, Vol. 29, No. 335, July 1982

90. Holtrop J., "A Statistical Re-analysis of resistance and Propulsion Data," International Shipbuilding Progress, Vol. 31, No. 363, November 1984

91. Oosterveld M.W.c., "Ducted Propeller Characteristics," Proceedings of Symposium on Ducted Propellers, Paper No.4, The Royal Institution of Naval Architects, 1973

92. Bingham V.P. and Mackey T.P., "High-Performance Rudders -with Particular Reference to the Schilling Rudder," Marine Technology, SNAME, Vol. 24, No.4, pp. 312-320, Oct. 1987

93. Dewey D.F. and Mitchell J.E., "Twin Schilling and 'High Lift' Rudders Frigate Model Manoeuvering Tests," Hamworthy Industramar Limited, Private Correspondence, U.K., June 1992.

94. van Baalen A.N., "A Flexible Offshore Craft Designed from Practical Experience," Reed's Tug World, Annual Review 1986/87.

95. Kletz T.A., HAZOP and HAZAN - Notes on the Identification and Assessment of Hazards, Institution of Chemical Engineers, 1986

96. Knowlton R.E., Hazard and Operability Studies and Their Initial Applications in R & D, R & D Management, Vol. 7, No.1, 1976

97. Oakland H., "Total Quality Management," Published by Heinemann, 1989

98. Jacobs W.R., "Estimation of Stability Derivatives and Indices of Various Ship Forms and Comparison with Experimental Results," Davidson Laboratory Report 1035, 1964

99. Jacobs W.R., "Method of Predicting Course Stability and Turning Qualities of Ships," Davidson Laboratory Report 945, March 1963

# 14 BIBLIOGRAPHY

2011 C G Biancardi et alii, "Towards Zero/Low Emissions Passenger Ferries," SNAME Met Section, NY, March 10th, 2011

2010 C G Biancardi, "The Hybridisation of power-train of ships: Principles and State of the Art Report", Annali Università Parthenope Napoli, Volume XX, 2010

2010 C G Biancardi, «Mobilité intra-rade et le projet DEESSE, Navette passagers à faible émission de GES », Salon Water Symposium, Cannes le 2 juillet 2010

2009 C G Biancardi, "When Luxury becomes Environ mental Respectful", Seatrade Miami, 2009

2007 C G Biancardi, "EIRAC: Visions of the intermodal transport: bottlenecks and research needs 2007-2020", http://www.itst2007.eurecom.fr/site/var/html/h1053/file1205.pdf

2006 C G Biancardi, Editorial Board: ICTE in Regional Development, ISBN 9984-633-03-9

2005 C G Biancardi et alii, "MULTI-MODAL TRANSPORT PRICING AND COSTING ANALYSES", REALISE GTC2-2000-33032 (Regional Action for Logistical Integration of Shipping across Europe), 2005

2005 C G Biancardi et alii, "Final Report on Economic Transport Performance", REALISE GTC2-2000-33032, Regional Action for Logistical Integration of Shipping across Europe, AMRIE 2005

2004 C G Biancardi, "Shipping quality and safety of High-Speed Vessels, terminals and Ports operations In Nodal points", EC DG TREN, 2004

2003 C G Biancardi, "SPIN-HSV: TOWARDS IMPROVED HIGH SPEED MARITIME TRANSPORT", The Naval Architect, RINA, April 2003

2002 Biancardi et alii, " SEAM - Measures to minimize environmental impacts from ships,", ENSUS conference, University of Newcastle, December 2002, Newcastle, UK

1997 C G Biancardi, "An alternative methodology for calculating ship maneuvering coefficients at the design stage", Delft University Press, ISSN: 0020-868X CODEN: ISBPAS, International shipbuilding progress A. 1997, vol. 44, n° 440, pp. 273-297 [bibl. : 17 ref.]

1996 Biancardi C G, Cavazzi D, Graziano G, Masullo MT, "Full scale measurements of a set of yaw/sway maneuvering Q-indices", Marine Simulation and Ship Manoeuvrability, Chilsett Editor, Balkerma, Rotterdam, NL 1996 Biancardi, C.G. Cavazzi, D. Masullo, M.T. Parente, C., "Regional integration of information and communication technologies for navigation safety in European waters", Position Location and Navigation Symposium, 1996., IEEE 1996, 22-26 Apr 1996, pages: 92 – 95, Atlanta, GA, ISBN: 0-7803-3085-4

1995 C G Biancardi, "Principles of a set maneuvering indices based on Sway/Yaw phase relations", Society of Naval Architects and Marine Engineers, Journal of ship research, ISSN 0022-4502, 1995, vol. 39, no2, pp. 117-122 (16 ref.)

1995 Biancardi C G, Santamaria R, and Troisi S, "Engineering Applications of Innovative Kinematics GPS for Monitoring Sea Trials of High Speed Vessels", Proceedings of III Symposium on High Speed Marine Vehicles, Naples, Italy

1994 Biancardi C G and Vultaggio M, "Monitoring of Maneuvering Performances", NAV 94, Rome, Italy

1994 Biancardi C. G., "Integrating ship maneuverability with safety", ISSN: 0081-1661, Transactions - Society of Naval Architects and Marine Engineers A. 1994, vol. 102

1994 Biancardi C G, and Dellwo (Commentator); Barr R. A. (Commentator) et alii, "Integrating ship maneuverability with safety. Discussion." Authors' closure, Society of Naval Architects and Marine Engineers, ISSN : 0081-1661, Transactions - Society of Naval Architects and Marine Engineers A. 1994, vol. 102, pp. 447-474 [bibl. : 3 p.]

1993 C G Biancardi, "Integrating Ship Manoeuvrability with Safety", MARSIM 93, St John's Canada, pp 33-47

1992 C G Biancardi, D R Dellwo, "Maneuvering Criteria at the Design Stage", PRADS92, Vol2, Ed Elsevier Applied Science, 1.247

1992 C G Biancardi, "Ship control advisory system", Marine Technology Society, ISSN : 0025-3324, Marine Technology Society journal A. 1992, vol. 26, n° 3, pp. 47-55 [bibl. : 10 ref.]

1992 Biancardi C G et alii, "A LOCAL COOPERATIVE TRAFFIC EXPERT SYSTEM," Proceedings of International Conference on VTS, Vancouver, Canada, 1992

1991 C G Biancardi, D R Dellwo, "Classification of ships by their maneuvering characteristics", Society of Naval Architects and Marine Engineers, ISSN: 0081-1661, Transactions - Society of Naval Architects and Marine Engineers A. 1991, vol. 99, pp. 205-219 [bibl. : dissem.]

1991 C. G. Biancardi, " On Board Piloting Support System", Computer Applications in the Automation of Shipyard Operation and Ship Design 1991: Rio de Janeiro, Brazil, 10-13 September, 1991 205-216

1991 Biancardi C G, "ON A MATHEMATICAL EXAMPLE OF PREDICTION IN SHIP CONTROL" - Annali della Facoltà di Scienze Nautiche, 1991 – Istituto Universitario Navale, Napoli, Italy

1991 Biancardi C G, "ASYMMETRIC FORCE OF SHIPS ENGAGED OFF THE CENTERLINE OF A CHANNEL" - Annali della Facoltà di Scienze Nautiche, Istituto Universitario Navale, Napoli, Italy

1990 Biancardi C G, "Land-attached naval architecture for development of touristic resorts", Proceedings of the 1990 Congress on Coastal and Marine Tourism: a symposium and workshop on balancing conservation and economic development : Honolulu, Hawaii, USA, 25-31 May 1990

1990 C G Biancardi, "Maritime Maneuvering Piloting Aid", 9th Ship Control Systems Symposium, 10-14 September 1990, Vol 4, Bethesda, Maryland, USA

1990 C G Biancardi, R Balestrieri, " Evaluative Procedures of Ship Hydrodynamic Parameters to be considered in the Vessel Traffic Control in Congested and/or Restricted Waterways", Vessel Traffic Services in the Mediterranean, Vol. 1, pp. 189-202 (1990)

1990 C G Biancardi, "Practical calculation method of ship maneuvering characteristics at the design stage", Delft University Press, ISSN : 0020- 868X, International shipbuilding progress A. 1990, vol. 37, n° 411, pp. 221-245 [25 pages] [bibl. : 22 ref.]

1990 C G Biancardi, S Romano, "Asymmetric hydrodynamic coefficients of ships engaged off the centre-line of a channel", Annali Istituto Universitario Navale, Volume L, 1990

1989 C G Biancardi, R Balestrieri, "Numerical Evaluation of Ship Maneuvering in Confined and Shallow waters", PRADS 1989, Varna, Bulgaria, 23-28 October, 1989

1989 C G Biancardi, R Balestrieri, S Mauro, "Verifica di un modello matematico semplificato per il controllo della traiettoria di una nave - Validation of a simplified mathematical model for ship control", Journal Tecnica Italiana, ISSN 0040-1846, 1989, vol. 54, no4, pp. 271-286 [16 page(s) (article)] (39 ref.)

1989 C G Biancardi, "Stability Analysis of Fire-Fighting, Considerations and Recommendations", PRADS 1989, Varna, Bulgaria, 23-28 October, 1989

1988 C G Biancardi, and Balestrieri, R , "Computational Hydrodynamics and Ship Maneuvering Characteristics, Proceedings of International Conference CADMO, 1988.

1988 C G Biancardi, "Applications of artificial intelligence in the management of marine operations", Bulletin of the Permanent International Association of Navigation Congresses A. 1988, n° 60, pp. 115-125 [bibl.: 60 ref.]

1988 C G Biancardi, D R Dellwo, "Machine management of marine operations", Society of Naval Architects and Marine Engineers, ISSN: 0025-3316, Marine technology A. 1988, vol. 25, n° 1, pp. 30-35 [bibl.: 59 ref.]

1988 C G Biancardi "On a Simplified Mathematical Model for an Onboard Maneuvering Simulator and its Potential Applications to the Ship Control", Invited Paper, Proceedings of Pacific Congress on Marine Science, Hawaii, USA, 1988

1988 C G Biancardi, "On a simplified mathematical model for an onboard maneuvering simulator and its potential applications to ship control", International shipbuilding progress A. 1988, vol. 35, n° 401, pp. 81-95 [bibl.: 29 ref.]

1988 C G Biancardi, "A simplified mathematical model for an onboard maneuvering simulator", SIMULATION, Vol. 51, No. 2, 75-81 (1988), DOI: 10.1177/003754978805100207

1988 C G Biancardi, "Privatization: a recent example at the National Maritime Research Center", Maritime Policy & Management, Volume 15, Issue 4 October 1988 , pages 275 - 278

1987 C G Biancardi, R Balestrieri, "Research Guidelines fro Developing and Onboard Maneuvering Simulator dor the ship control" Annali Istituto Universitario Navale, Volume LVI, 1987-1988

1987 C G Biancardi, M Capecchi, G Mallardo, A Troiano, A Trotta, "Sistema Consultivo per il Pilotaggio in Sicurezza di una Nave" Annali Istituto Universitario Navale, Volume LVI, 1987-1988

1985 C G Biancardi, R Balestrieri, N Petronzi "A simplified model of human error in the manouevering of ships, Annali Istituto Universitario Navale, Volume LIV, 1985

1985 C G Biancardi, "Considerations on Thesis and Results discussed at MARISM 84, Annali Istituto Universitario Navale, Volume LIV, 1985

1985 C G Biancardi, Balestrieri, "L'errore umano nella manovra della nave", Studi Marittimi, Anno VIII, N 25, June 1985

1984 C G Biancardi, "Le indicazioni emerse dal Marine Engineering Seminar nel campo della tecnica navale", Annali

Istituto Universitario Navale, Volume LIII, 1984

BOOKS and Chapters in BOOKS

1996 Biancardi C G, Cavazzi D, Graziano G, Masullo MT, "Full scale measurements of a set of yaw/sway maneuvering Q-indices", Marine Simulation and Ship Manoeuvrability, Chilsett Editor, Balkerma, Rotterdam, NL

1990 C G Biancardi, Book "Principles of Safety of Marine Vehicles", Istituto Universitario Navale, Italy

1988 C G Biancardi and R Balestrieri, "Computational Hydrodynamics and Ship Maneuvering", in the Book "Marine and Offshore Computer Applications", Springer-Verlag, 1988

1986 C G Biancardi, Il Simulatore di Manovra Navale come strumento didattico ed addestrativo, in book "Corso di aggiornamento in Automazione Navale per docenti degli Istituti Nautici, Italia, 1986

# 15 APPENDICES

APPENDIX 1
A METHOD OF CLASSIFYING THE SHIPS' MANOEUVRING CHARACTERISTICS

## 1.1. Introduction

Chapter 7 of the book has presented alternative maneuvering criteria based on fresh indices of maneuverability. The unit standard deviation criteria of Chapter 7 provide a natural statistical means for classifying hulls in a database. Conventional bounds for $Q_H'$, $Q_{HR}'$, $Q_R'$ at zero trim are calculated from the database, see Chapter 7, as follows:

$$-.331 \leq Q_{H'} \leq 0 \tag{1.10a}$$

is a unit standard deviation criterion for acceptable bare-hulls at zero trim, and

$$Q_{H'} < -.331 \tag{1.10b}$$

is the criterion for unacceptable hulls. And,

$$-.331 \leq Q_{H'R} \leq 0 \tag{1.11a}$$

for acceptable turning behavior during transit to a steady straight course, and

$$Q_{H'R} < -.331 \tag{1.11b}$$

for unacceptable behavior. While,

$$-.145 \leq Q_{R'} \leq 0 \text{ for acceptable hull-rudders} \tag{1.12a}$$

$$Q_{R'} < -.145 \text{ for unacceptable hull-rudders} \tag{1.12b}$$

provides a conventional rudder performance criterion when exiting from a steady straight course at zero trim.

However, (1.10), (1.11) and (1.12) require computation of the indices $Q_H'$, $Q_{HR}'$ and $Q_R'$ and this makes them poor design tools. First, the indices are not easily formulated in terms of ship geometry. Second, the ship designer is usually not interested in the value of the indices but rather in the assurance that (1.10), (1.11) and (1.12) are satisfied. Consequently, it would desirable to develop alternative rules that are equivalent or nearly equivalent to the three criteria but do not involve the indices directly and are more convenient to use than (1.10), (1.11) and (1.12).

Each criterion is used to partition the database scatter diagram into two distinct clusters: a normal group of ships and an abnormal one.

The main objective of this Appendix is to use statistical methods, [79], to identify lines separating the clusters. The result is a method of classifying ships' maneuvering characteristics based on a rule that assigns a ship or ship design to the group with highest condition probability. The rules derived in this way are linear expressions formulated in terms of basic design parameters. They can be applied without evaluating the indices. Most important, they can be readily used to assess design during either steady or transient motion.

## 1.2. Maneuvering Classification rules.

The unit standard deviation (1.10) provides a natural statistical means for classifying bare-hulls in the database. However, (1.10) requires computation of the index $Q_H'$. It would be desirable to develop an alternate rule that is equivalent or nearly equivalent to (1.10) but does not involve the index directly.

A rule of this type can be derived by plotting the index $Q_H'$ against the hull's slenderness. The variation of $Q_H'$ with respect to $S_1$ is shown in Figure 1.1. Hulls in the database that satisfy (1.10a) are represented by

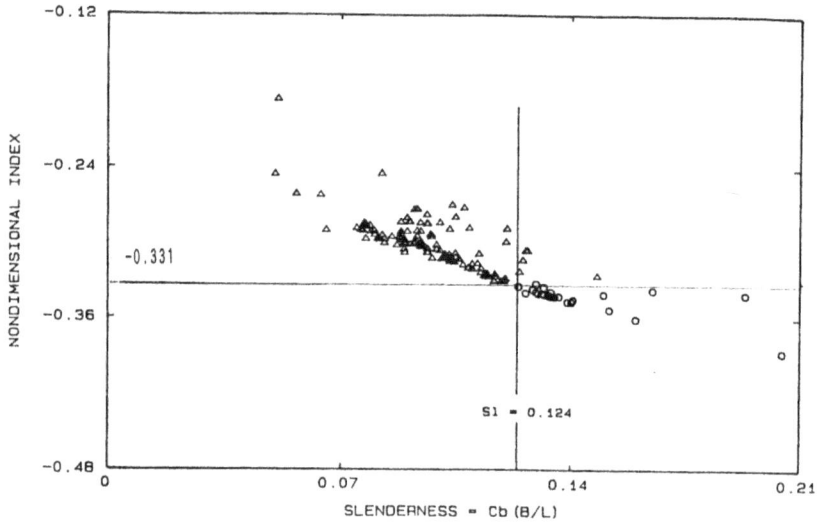

Fig. 1.1

$Q_H'$ plotted against $S_1$ for each ship in the database.

Normal hulls that satisfy (1.10a) are indicated by triangles, while abnormal hulls satisfying (1.10b) are indicated by circles.

triangles above the index level $Q_H'=-.331$. Hulls that satisfy (1.10b) are represented by circles below that level. Conditions (1.10a) and (1.10b) partition the database into easily recognizable groups divided by the horizontal line at $Q_H'=-.331$. Inspection of the figure indicates that virtually the same partition can be achieved by drawing a vertical line at $S_1=0.124$, where $S_1$ is the slenderness variable. Thus

$S_1 \leq 0.124$ for normal hulls                                (2.10a)

$S_1 > 0.124$ for abnormal hulls                            (2.10b)

is a single variable linear classification rule that is nearly equivalent to (1.10) but does not require computation of the index.

It is evident from Figure 1.1 that (2.10) misclassifies 5 of the 173 hulls, an error rate of 2.9%.

Condition (2.10a) and (2,10b) provide the designer with an elementary means of relating hull performance to the slenderness parameter. Often, however, the ship designer would prefer to relate performance to the components of slenderness, cb and B/L. Unfortunately, it is not possible to derive accurate single-variable linear classification rules analogous to (1.10) using these variables. However, it is possible to derive an accurate two-variable rule that involves cb and B/L.

The high accuracy of (1.10) is a consequence of the non-layered quality of the distribution in Figure 1.1.
When the distribution is projected onto the slenderness axis, the clusters of normal and abnormal hulls remain distinct. In contrast, when the layered distributions shown in Figures 1.2 and 1.3 are projected onto their horizontal axes, the clusters of normal and abnormal hulls mix and lose their separate identities. Figure 1.2 is a sketch of $Q_H'$ against the hull's block coefficient and Figure 1.3 is a sketch of $Q_H'$ against the aspect ratio B/L. The figures indicate that it is not possible to replace (1.10) with a single variable rule based on either cb or B/L that is as accurate as (1.10).

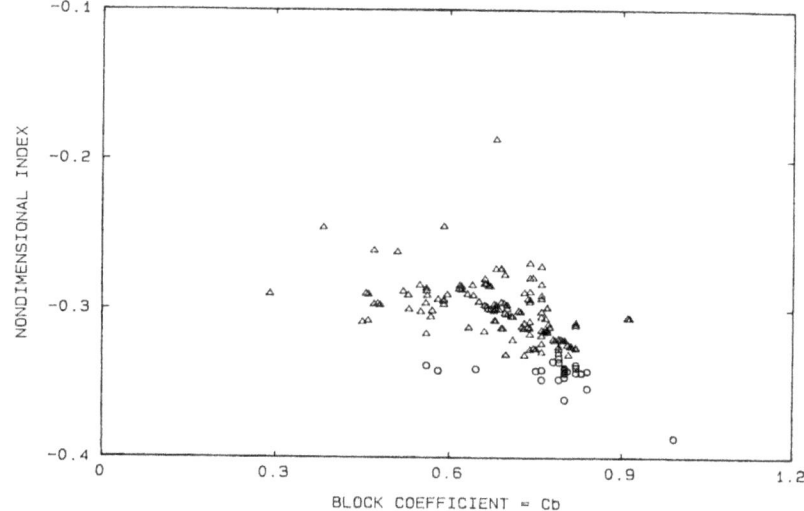

Fig. 1.2

$Q_H'$ plotted against cb for each ship in the database.

Normal hulls that satisfy (1 .10a) are indicated by triangles, while abnormal hulls satisfying (1.10b) are indicated by circles.

A two-variable classification rule can be derived by projecting a three-dimensional scatter diagram of the index distribution onto the two-dimensional design space spanned by cb and B/L. The projected distribution is shown in Figure 1.4. Inspection of the figure indicates

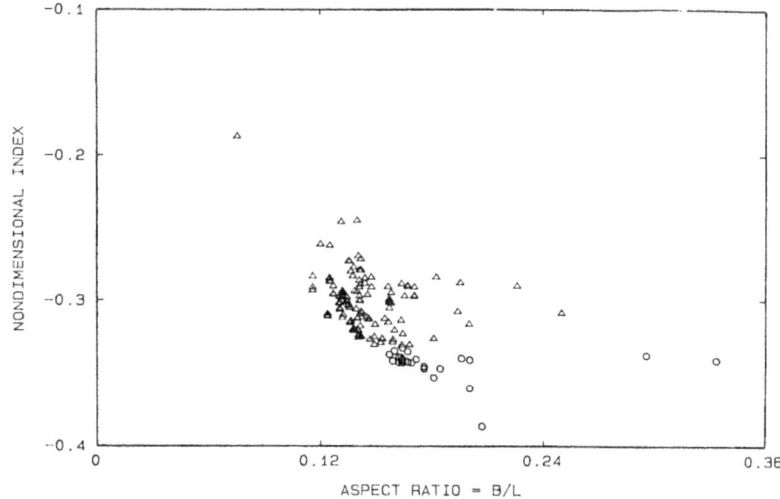

Fig. 1.3

$Q_H'$ plotted against B/L for each ship in the database.

Normal hulls that satisfy (1.10a) are indicated by triangles, while abnormal hulls satisfying (1.10b) are indicated by circles.

that the normal and abnormal hulls form clear and distinct groups. It is important to realize that while this clustering occurs in the two-dimensional design space of Figure 1.4, it does not occur in either of the one-dimensional design spaces that form the horizontal axes in Figures 1.2 and 1.3.

The line appearing in Figure 1.4 is given by

$$W_H \text{ (cb, B/L)} = \ln(p_2/p_1) \tag{2.11}$$

where

$$W_H(cb, B/L) = 20.561cb + 103.505(B/L) - 31.996 \tag{2.12}$$

and $p_1 = 0.37$ and $p_2 = 0.63$ are a priori probabilities. The classification line is the one-dimensional analogue in cb x (B/L) space of the isolated point $S_1 = 0.124$ on the slenderness axis in Figure 1.1. The line and point discriminate between normal and abnormal hulls. The linear expression (2.12) defining the classification line was developed from the database using the statistical procedure in [79]. A point (cb, B/L) in a design space that lies on (2.11) is as likely to correspond to a hull with an index satisfying (1.10a) as it is to a hull whose index satisfies (1.10b). A point that lies below the discrimination line in Figure 1.4 is more likely to represent a hull with $-0.331 \leq Q_H' \leq 0$ than a hull with $Q_H' < -0.331$. Thus (2.11) leads to the classification rule

$$W_H \text{ (cb, B/L)} \leq \ln(p_2/p_1) \text{ for normal hulls} \tag{2.13a}$$
$$W_H \text{ (cb, B/L)} > \ln(p_2/p_1) \text{ for abnormal hulls} \tag{2.13a}$$

or

$$W_H \leq .532 \text{ for acceptable bare hulls} \tag{2.13c}$$
$$W_H > .532 \text{ for unacceptable bare hulls} \tag{2.13d}$$

This is a linear rule involving two variables that is nearly equivalent to (1.10) but does not requires computation of $Q_H'$. The rule's error rate of 2.6% is comparable to that of (2.10). It classifies all of the abnormal hulls satisfying correctly (1.10b) correctly. It misclassifies 4.1% of the normal hulls.

Similarly, a two-variable classification rule can be derived by projecting a three-dimensional scatter diagram of the index distribution onto the two-dimensional design space spanned by $S_1$ and $\sqrt{(Ra)}/D$. That office-ready rule alternative to (1.11) is given by

$$W_{HR} \leq .444 \tag{2.14a}$$
for acceptable hull-rudder combinations;

$$W_{HR} > .444 \tag{2.14b}$$
for unacceptable hull-rudder combinations;

here:

$$W_{HR}(S_1, \sqrt{(Ra)}/D) = 125.505(S_1) - 2.951(\sqrt{(Ra)}/D - 13.035$$

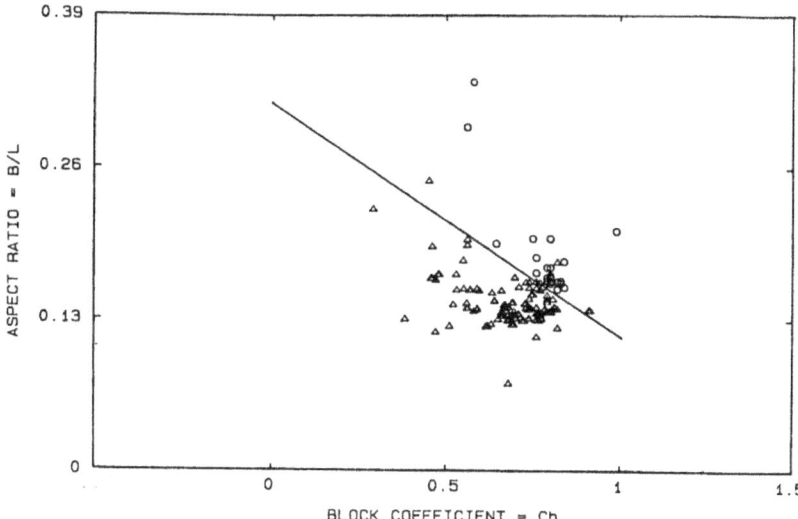

Fig. 1.4

A scatter diagram in the design space spanned by Cb and B/L. Each point represents a ship in the database. Normal hulls are represented by triangles and abnormal hull are represented by circles.

The rule partitions the design space of Figure 1.5 into two regions: the region above approach criterion line ($W_{HR}$ = .444) and the region below that line. Points ($S_1$, $\sqrt{(Ra)}/D$) that lie above the criterion line correspond to unacceptable hull-rudder designs. Points that lie below the criterion line correspond to hull-rudder designs that are expected to be well behaved ($-.331 \leq Q_{HR} \leq 0$) on approach to a steady straight course. The rule misclassifies 4% of the hull-rudder combinations studied.

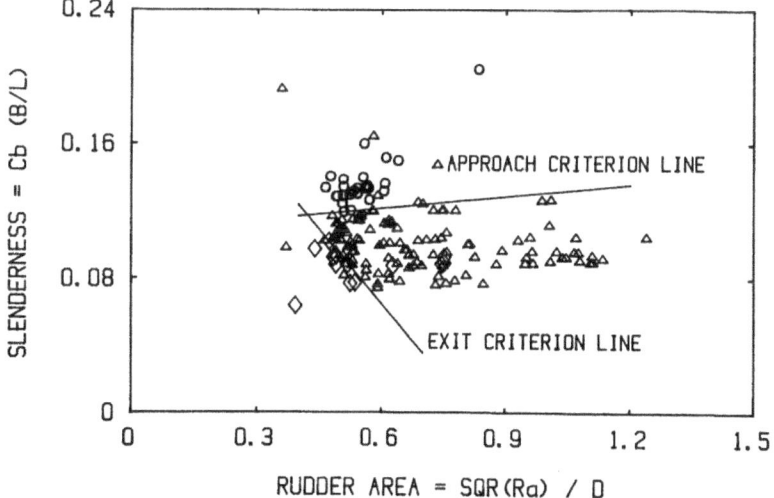

Fig. 1.5

Each point represents a ship in the database. Triangles represent normal ships, whose hull-rudder index satisfies (1.11a) and whose rudder index satisfies (1.12a). Ships that exhibit abnormal or unacceptable behavior are represented by circles and diamonds. Circles correspond to ships whose hull-rudder index satisfies (1.11b). Diamonds correspond to ships whose rudder index satisfies (1.12b).

Finally,
$$W_R \leq 20.546 \qquad\qquad\qquad (2.15a)$$

for acceptable rudder performance;

$W_R > 20.546$ (2.15b)

for unacceptable rudder performance.

is a nearly equivalent practical alternative to the exit criterion (1.13). Here:

$W_R (S_1, \sqrt{(Ra)}/D) = -14.366(S_1) - 4.2171(\sqrt{(Ra)}/D) + 24.005$

defines a linear classification rule that part ions the design space of Figure 1.5 into two regions. The region below the exit criterion line ($W_R = 20.546$) corresponds to ships that are expected to exhibit unacceptable behavior ($Q_R' < -.145$) on exit from a steady straight course. The region above the criterion line corresponds to ships with acceptable exiting behavior ($-.145 \leq Q_R' \leq 0$). Although this rule correctly classifies 95% of the ships in the database, it is not as accurate as rules (2.13) and (2.14). It misclassifies 29% of the abnormal ships. The large error is due to the fact, mentioned earlier in Chapter 7, that the rule is based on only 82 of the 173 ships in the database.

## 1.3. Conclusions

Based on the results of this Appendix, the following conclusions can be drawn:

a) Linear classification rules can be employed to distinguish between ships with acceptable maneuvering qualities and ships with unacceptable qualities avoiding the calculation of the indices.

b) The classification rules involve one or more ship geometric parameters and can be used to satisfy the maneuvering criteria of Chapter 7 at the design stage.

APPENDIX 2
A METHOD FOR PREDICTING SHIP MANOEUVRING PERFORMANCES

## 2.1. Introduction

At the design stage, prediction of ship maneuvering performances can be achieved by solving the Newton's equations of motion in the horizontal plane, see [15].

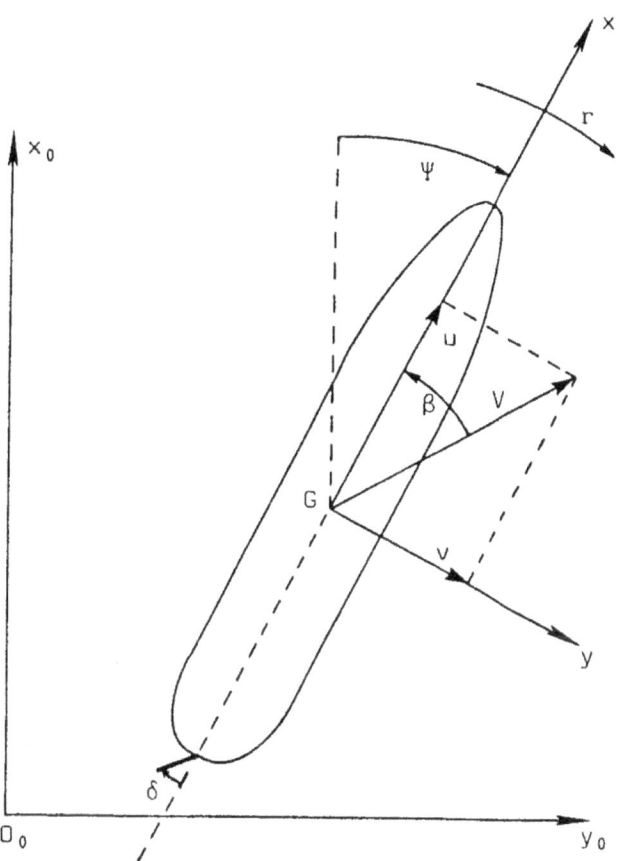

Fig. 2.1.
Body-fixed and space-fixed coordinate systems. The body–fixed system is located at the ship's center of gravity G

Basic equations in the horizontal plane can be considered first with reference to one set of axes fixed relative to the earth and a second set fixed relative to the ship. In spite of the apparent simplicity of the equations, the motion of a ship is more conveniently expressed when referred to the axes fixed respect to the moving ships, see Figure 2.1. The differential equations of motion in the horizontal plane are:

$(m+m_x)u - (m+m_y)vr = X$ ⠀⠀⠀⠀(Surge)

$(m+m_y)v - Y_r r + (m+m_x)ur = Y$ ⠀⠀(Sway) (2. 1 )

$(I_z+J_z)r - N_v v = N$ ⠀⠀⠀⠀⠀(Yaw)

The terms in the equations are shown in Figure 2.1 and explained in Appendix 10.

For ship-like body of revolution, the hydrodynamic forces X and Y and moment N of the equations of motion are normally described by Taylor series about the equilibrium condition ($v=r=\delta=0$), resulting in different terms (generally called hydrodynamic coefficients or derivatives) necessary to describe the ship dynamics and removing those terms that are zero for bodies with a vertical plane of symmetry, and ignoring terms of fifth and higher order, as follows, (for symbols see Appendix 10) :

$$X' = \frac{X}{\Omega/2 \, AxV} = \qquad\qquad (2.2a)$$

$$= X_{rr}'r'^2 + X_{vr}'v'r' + X_{vv}'v'^2 + X_{\delta\delta}\delta^2$$

$$Y' = \frac{Y}{\Omega/2 \, AxV} = \qquad\qquad (2.2b)$$

$$= Y_v'v' + Y_r'r' + Y_{vr}'v'r'^2 + Y_{rvv}'r'v'^2 + Y_{vvv}'v'^3 + Y_{rrr}'r'^3 + Y_\delta\delta + Y_{\delta\delta\delta}\delta^3$$

$$N' = \frac{N}{\Omega/2 \, AxV} = \qquad\qquad (2.2c)$$

$$= N_v'v' + N_r'r' + N_{vr}'v'r'^2 + N_{rvv}'r'v'^2 + N_{vvv}'v'^3 + N_{rrr}'r'^3 + N_\delta\delta + N_{\delta\delta\delta}\delta^3$$

The evaluation of the ship responses in surge-sway-yaw requires the calculation of the hydrodynamic forces acting on the hull, propeller and other appendages and their interactions along X, Y-axis, and corresponding moments of rotation around Z-axis. This can be achieved using various techniques for calculating the hydrodynamic coefficients in (2.2). Indeed, the hull hydrodynamic forces, in a turn, are separated into two large categories:

a) Linear and Quadratic Coefficients account primarily for viscous dissipative losses of a ship in a maneuver.
These terms are important in all stages of a maneuver. Viscosity affects the flow around the hull in at least two ways. At small drift angles, the ship hull and the appendages such as fins, bilge keels and sonar domes, develop a circulating cross force, and therefore, can be regarded as a lifting surface with the drift angle analogous to the conventional rudder angle of attack. The appropriate coefficients $Y_v'$, $Y_r'$, $N_v'$ and $N_r'$ are constant factors of the linearly dependent sway and yaw velocities and damping forces. At larger drift angles, the ship hull and the appendage damping forces are generated by pressure loss and, therefore, contribute to the losses in oblique flow with drift and yaw angle velocities by cross-flow pressure. The hydrodynamic coefficients are $X_{rr}'$, $X_{vr}'$, $X_{vv}'$, $Y_{vrr}'$, $Y_{rvv}'$, $Y_{vvv}'$, $Y_{rrr}'$, $N_{vrr}'$, $N_{rvv}'$, $N_{vvv}'$, $N_{rrr}'$. In typical maneuvers, both circulatory and cross-flow forces are present and equally important.

b) Inertial Hydrodynamic Forces associated with the "added mass" constant terms and accelerations in surge, sway and yaw motions. The corresponding linear no dimensional coefficients are functions of ship hull geometry and location of appendages. These terms are important during the initial and transient part of a maneuver. While, the contribution of these terms in the steady part of a turn is small. Because the inertial forces are net greatly influenced by the viscous effects, the computational methods based on the potential, inviscid flow can also be used for ship maneuvering assessment.

The resistance and propulsion contributions to X, Y and N can be based, with reasonable accuracy, on statistical analysis of a large number of model tests.

The calculation of the coefficients in (2.2a) can estimated with acceptable accuracy using the calculation methods described in [22]. However, at the present time, there are not analytical methods calculate accurately the coefficients in (2.2b) and (2.2c), and the most reliable method is still by captive model testing.

At the design stage, it would be useful if the other hydrodynamic coefficients acting on the ship, can be estimated from the main geometric parameters of a ship's body form. In the past, methods based on geometric parameters such as L, B, T and cb have been proposed, [9, 11, 41]. However, these methods do not take always into consideration changes of the ship form, for example changes of the stern or bow design, or type or location or rudder and propeller.

The main objective of this Appendix is to calculate the hydrodynamic coefficients in (2.2b) and (2.2c) of surface ships by applying adjusted formula obtained previously for submerged bodies, [80], and that:

• take into account the ship form and geometric relationships;

• include the free surface effects; and

• consider the effects of the propeller-hull-rudder interaction.

In Appendices 4, the results are verified and correlated by comparison with model test data of surface ships.

## 2.2. Method for Calculating the Linear and Quadratic Hydrodynamic Coefficients

<u>Assumptions</u>. In [80], a method for predicting the hydrodynamic coefficients of submersibles is presented.
The formulae presented in [80] for the hydrodynamic coefficients of deeply submerged bodies are based upon theory modified by experimental results. In the case of submersibles, the absence of the free surface effect has permitted to formulate each hydrodynamic coefficient of (2.2) in terms of the geometric parameters of the submerged bodies which consists of a body of revolution, tail fins with movable rudders and stern propellers, [80]. However, for surface displacement ships, adequate formulae for the hydrodynamic coefficients of the side force Y and yaw moment N, in terms of the vessel geometry have not been previously fully developed. Despite the presence of the free surface, model experiments, [8], show that at low Froude numbers (Fn) at which conventional surface ships commonly operate, the hydrodynamic coefficients are not significantly dependent on forward speed.

Even though the differences in section shapes of surface ships and submersible hulls must be taken into accounts, tests of submersible hulls with circular [81] and rectangular cross sections [82] of surface bodies exhibit marked similarities in side force Y and yaw moment N characteristics, especially when the draft-to-beam ratio of the rectangular hull is small (as it is for the surface ship).

Thus, it is possible to use previous formula developed for submersibles in [80], opportunely modified for surface effects, for calculating the hydrodynamic coefficients of ships, within the following assumptions:

a) Low Froude Number;

b) Draft-to-beam ratio of hull is small;

c) The surface ship can be represented as a combination of a body of revolution, a distribution of fixed and movable fin areas, and propellers similarly to the submersibles configurations;

d) The choices of the body of revolution and fin areas used to represent the ship must be uniquely determined for each ship form;

e) The fin portions of the hull are defined by its "thin" sections which are <u>tentatively</u> taken to mean those sections where the hull thickness $t_h$ is less than one-tenth of the maximum beam B.

<u>Hydrodynamic Coefficients of Surface Ships</u>. The body plans and profile drawings are used to determine the profile shapes of the wetted hull areas where $t_h \leq 0.1 \bullet B$, and these areas are treated as fins. The fin area often will consist of a large triangular or rectangular area plus a long, narrow region of vanishingly small area, because of the nature of the merchant ship hull form, as indicated in the Figure 2.2. However, because the $t_h$ criterion is arbitrary, the sensitivity of the predicted hydrodynamic characteristics to changes in the definition of "thin" should be investigated. It can be expected on the basis of the formulae for the fin effects that this sensitivity will be low as long as $t_h$ is much smaller than the maximum beam B. For symbols see Appendix 10.

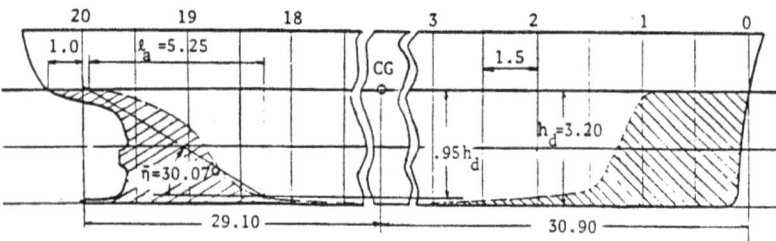

Fig. 2.2.
Body plan, profile and particulars of a sample hull. In gray the fin sections.

<u>a. Calculation of fin aspect ratio</u>
ARtf = 2· htf² / (Atf + Ati)
ARbf = 2· hbf² / (Abf + Abi)

In ARbf, the factor 2 appears because at very low Froude numbers the free water surface, acts Like a wall.
Moreover, because of the nature of most surface ship hull forms, the fin area often will consist of a large, triangular or rectangular area plus a long, narrow region of vanishingly small area, as indicated in Figure 2.2, it is difficult to define the included hull area Abi. Such a difficulty can be avoided if Abi is defined by choosing Abi=Abf, so that

ARbf = hbf² / Abf

<u>b. Calculation of the effective base area hull (Abe)</u>
θ = ATN (.95-T - hb) / la) hull afterbody angle with hb = height of the base of the solid of revolution usually for conventional hulls hb = 0
If la = 0 then θ = 0

Abe= (.55+.45 ·**Ab/Ax**)-(1.712 · **θ**-2.41 · $\theta^2$+√| .637* **θ**|)+ .725 · $\theta^3$) · **(1-Ab/Ax)**

<u>c. Calculation of the bow draft (htf) and the tail draft (hbf)</u>. htf and hbf are measured directly from the drawings of the ship design. However, in the case that the bow fin area is a long and thin section htf is calculated as:

htf = 4·Atf· TAN(θ)

and ARtf becomes ARtf = 4· TAN(θ)

<u>d. Linear Coefficients</u>
Yv = -(Yb + Ybtf + Ybbf + Ybr + Ybp + Ybs)
where:

COMPONENTS

Yb=2.13 Abe                                                     base

If ARtf is >1.85 but ≤ 6 then
Ybtf = (2-1/12·ARtf)·Atf/Ax                           tail fin

If ARtf is > 0 but ≤ 1.85  then
Ybtf = 2 ARtf·Atf/Ax                                       tail fin

Ybbf= (1.8 · pig· ARtbf·Abf/Ax)/( √(Arbf²+4)+1.8))     bow fin

Ybr= (2-1/12·Arr) ·Ar/Ax                                    rudder

Ybp=3.7·ARp·(Ap/Ax)(3+ARp)                        propeller

If ARs is >0 but ≤ .185 then
**Ybs= 2·ARs·Ass/Ax**                                        skeg

If ARs is >1.85 but ≤ 6 then
Ybs = (2-1/12·ARs)·Ass/Ax                          skeg


Nv= - (Nbi + Nb + Nbtf + Nbbf + Nbr + Nbp + Nbs)
Yr= - (Yrb + Yrtf + Yrbf + Yrru + Yrp + Yrs)
Nr= - (Nrb + Nrtf + Nrbf + Nrru + Nrp + Nrs)

where:
                                                         COMPONENTS
Nbi =1.9 · cp · (k 2 · k1)                                 ideal

Nb=xbe· Yb                                                    base

Yrb=xbe·Yb
Nrb=xbe$^2$ ·Yb

Nbtf=xtf·Ybtf                                             tail fin
Yrtf=xtf·Ybtf
Nrtf=xtf$^2$·Ybtf

Nbbf=xbf·Ybbf                                             bow fin

Yrbf=xbf·Ybbf
Nrbf=xbf$^2$·Ybbf

Nbr=xr·Ybr                                                rudder
Yrru=xr·Ybr
Nrru=xr$^2$·Ybr

Nbp=xp·Ybp                                               propeller
Yrp=xp·Ybp
Nrp=xp 2·Ybp

Nbs=xs·Ybs                                                  skeg
Yrs=xs·Ybs
Nrs=xs$^2$ ·Ybs

e. Nonlinear Coefficients. Third order normal and side force coefficients. Indicating Cdc as the steady-state drag coefficient of an infinite cylinder in a flow normal to its axis, it can be shown that a factor "n" represents the reduction in Cdc due to the finite length of the body and that "$ne_0$" and "$ne_1$" represent the reduction of Cdc for a hull-like form, [83J.
$ne_0$ ≈.21
$ne_1$ ≈. 1

$Yvrr = b3b + b3f$
$Yrvv = b4b + b4f$
$Yvvv = b5b + b5f$
$Yrrr = b6b + b6f$

$Nvrr = f3b + f3f$
$Nrvv = f4b + f4f$
$Nvvv = f5b + f5f$
$Nrrr = f6b + f6f$

where:
$b3f = -12 \cdot (btf2 + bbf2 + bs2 + br2 + bp2)$
$b4f = -12 \cdot (btf1 + bbf1 + bs1 + br1 + bp1)$
$b5f = -4 \cdot (btf + bbf + bs + br + bp)$
$b6f = -4 \cdot (btf3 + bbf3 + bs3 + br3 + bp3)$

$f3f = -12 \cdot (btf3 + bbf3 + bs3 + br3 + bp3)$
$f4f = -12 \cdot (btf2 + bbf2 + bs2 + br2 + bp2)$
$f5f = -4 \cdot (btf1 + bbf1 + bs1 + br1 + bp1)$
$f6f = -4 \ 0 (btf4 + bbf4 + bs4 + br4 + bp4)$

$xm = L/2 - xcg$

$b3b = -.98 \cdot ne_0 \cdot la \cdot (3/2 \cdot xm^2 - 2 \cdot xm + 3/4)$
$b4b = -.98 \cdot ne_1 \cdot la \cdot (3/2 \cdot xm - 1)$
$b5b = -.49 \cdot ne_0 \cdot la$
$b6b = -.98 \cdot ne_0 \cdot la \cdot (xm^{(3/2)} - xm^2 + 3/4 \cdot xm - 1/5)$

$f3b = -.98 \cdot ne_0 \cdot la \cdot (3/2 \cdot xm^2 \cdot 9/4 \cdot xm - 3/5)$
$f4b = -2 -.98 \cdot ne_0 \cdot la \cdot (3/2 \cdot xm^2 - 2 \cdot xm + 3/4)$
$f5b = -.98 \cdot ne_0 \cdot la \cdot (xm/2 - 1/3)$
$f6b = -.98 \cdot ne_0 \cdot llaa \cdot (xm^4/2 - 4/3 \cdot xm^3 + 3/2 \cdot xm^2 - 4/5 \cdot xm + 1/6)$

$Atf = Atf / aaxx$                                    tail fin
$btf = Atf / ARtf$
$btf1 = btf \cdot xtf$
$btf2 = btf \cdot xtf^2$
$btf3 = btf \cdot xtf^3$
$btf4 = btf \cdot xtf^4$

$Abf = Abf / aaxx$                                    bow fin
$bbf = Abf / ARbf$
$bbf1 = bbf \cdot xbf$
$bbf2 = bbf \cdot xbf^2$
$bbf3 = bbf \cdot xbf^3$
$bbf4 = bbf \cdot xbf^4$

$Ass = Ass / aaxx$                                    skeg
If $ARs > 0$ then $bs = Ass / ARs$
$bs1 = bs \cdot xs$
$bs2 = bs \cdot xs^2$
$bs3 = bs \cdot xs^3$
$bs4 = bs \cdot xs^4$

Ar = Ar / aaxx                                                              rudder
If Ar > 0 then br = Ar / ARr
br1 = br · xr
br2 = br · xr$^2$
br3 = br · xr$^3$
br4 = br · xr$^4$

Ap = Ap / aaxx                                                             propeller
Ap -= 2 ° Ap
If Ap > 0 thep
bp = Ap / ARp
bp1 = bp · xp
bp2 = bp · xp$^2$
hp3 = bp · xp$^3$
bp4 = bp · xp$^4$

## 2.3. Estimation of the Inertia ("Added Mass") Hydrodynamic Forces

The major added mass coefficients of the bare hull,
Yv=-m$_y$ and Nr=-Jz, are estimated using strip-wise methods or slender body approach based on the integration of the sectional added masses m$_y$(sectional)=kks · $\pi$· T$^2$/2, along the hull with the corresponding three-dimensional correction factors, [22]. Where kks is the Lewis form correction factor estimated according to reference [84, 85]

Yv = yvdo + yvdsk + yvdru + yvdpr

Nr = nrdo + nrdsk + nrdru + nrdpr

Where

yvdo= -$\pi$ ·kks·rr1·(T/L)$^2$
nrdo= -$\pi$/12 ·kks·rr2·(T/L)$^2$
rr1= 1- (.5·B+T)/(2·L)
rr2= 1- (.5·B+T)/(L)
If cm≥.785 then aass= 1.1·cm
If cm <.785 then aass=.9·cm

aass is the local sectional area of the corresponding Lewis form.

kks = 1 + aass · tanh(cm-.785)*B/T

a. skegs components
yvdsk = -((4· pig· hs/L) / ($\sqrt{}$(1 + ARs$^2$))) · Ass/L$^2$
nrdsk = xs$^2$ · yvdsk

b. rudders components
yvdru = 4 · (-((4· $\pi$· hr/L)/( $\sqrt{}$ (1+ARr$^2$))) · Ar/L$^2$)
nrdru = xr$^2$ · yvdru

c. propellers components
yvdpr = 2· (-((4· $\pi$· dp/L)/( $\sqrt{}$ (1+ARp $^2$))) · Ap/L$^2$)

$$nrdpr = xs^2 \cdot \ yvdsk$$

## 2.4. Conclusions

Based on the approach illustrated in this Appendix the following conclusions can be drawn:

a) At the design stage, an accurate and immediate prediction of ship maneuvering performance requires a calculation method of the hydrodynamic coefficients that takes into account the hull geometric parameters.

b) The ship hull can be represented as a combination of a body of revolution, a distribution of fixed and movable fin areas, and propellers;

c) The hydrodynamic coefficients of surface ships are computed by applying adjusted formula obtained previously for submerged bodies.

APPENDIX 3
A SIMPLIFIED METHOD OF CONSIDERING HULL DESIGN CHANGES IN THE CALCULATION OF MANOEUVRING PERFORMANCES

## 3.1. Introduction

At the design stage, many parameters that define the hull form still remains subject to change. These parameters are critical elements of the hydrodynamic coefficients that define the inherent maneuvering capability of the ship. In fact, different hull forms are likely to have different inherent maneuverability. For these reasons, before estimation can be made of the ship maneuvering capability, it is necessary to have a method for considering changes of an approximate hull form. In the past, other authors have presented methods based on use of polynomials, [86], to draw ship hulls. However, previous methods were directed towards ship surface design which is not required in this case.

By coupling the calculation method of the hydrodynamic coefficients of Appendix 2 with a geometric modeling method for varying the hull form, it is possible to investigate inherent ship maneuverability of planning ship designs by the use of computer techniques. For example, once fixed some main design parameters such as L, B, T, cb, cp, Ax, it is possible to consider the effect on the inherent maneuverability of different sets of sterns as well as various sets of rudders and propellers.

The main objective of this Appendix is to introduce a fresh method for approximating and changing hull form and configuration when fixing some of the design parameters. The method has to provide input data for the calculation method of the hydrodynamic coefficients of Appendix 2. In this book, the hull is generated as a combination of two main polynomials representing the waterline and the keel-line. Moreover, because the inertia coefficients are easily calculated when each transverse section of the hull is approximated with a Lewis form, [84], each transverse section is conformed to a Lewis form. Input data of the method are the main geometric hull parameters such as L, B, t, cb. Output data of the method are the parameters, such as Ab, Abf, Abi, Af, Atf, Ati, cp, cm, and so on, needed for the calculation of the hydrodynamic coefficients illustrated in Appendix 2.

## 3.2. Description of the Method
Once some of design parameters are fixed, e.g. L, B, T and cb, cp, Ax, the hull is subdivided in five longitudinal parts. Along the keel, the variable positions of four key points (p1, p2, p3, pk) are used to vary the dimensions of five longitudinal parts. Figure 3.1 illustrates the variable locations of the key points.
Moreover, the hull is divided in a number of transverse sections that has a specific contour because of the hull form. For example, transverse sections contours at the bow-zone are likely to be different from the stern-zone ones.

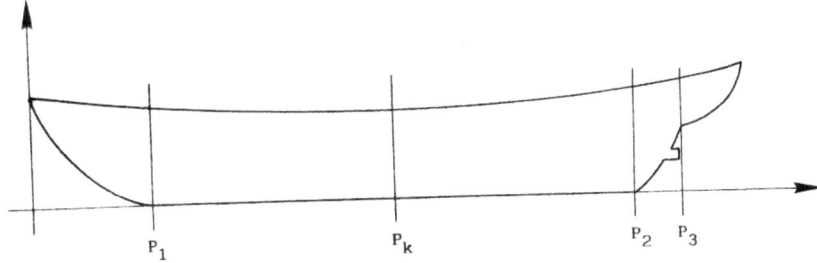

Fig. 3.1
The subdivision of the hull.

Integrating along the ship longitudinal axis, the volume, the midships sectional area, the mass, the added mass, the wetted hull surface and the mass moments of inertia are calculated.
Polynomials are used to calculate the varying maximum draft and beam of single transverse sections, as follows:

$$\text{tvmax} = T \cdot (1 - |x1/LL|)^k \cdot (1 - |xb/LL|)^q$$
$$\text{bvmax} = B/2 \cdot (1 - |xp/LLb|)^n \cdot (1 - |xb/LLb|)^m$$

where:
tvmax and bvmax are the maximum value of respectively draft and beam at the local section.

k, q, n, m are constant for each hull or part of hull
and x1=0 if lv $\leq$ L· p²

x1=L· p3-lv and LL=L· p3-L if lv $\leq$ L· p3
x1=L· p1-lv and LL=L· p1 if lv $\leq$ L· p1
x1=L· p2-lv and LL=L· p2 -l if L· p2 $\leq$ lv $\leq$ L· p3
xb=L· pk-lv
LLb=L· pk if lv $\leq$ L· pk
LLb=L· pk-L if lv > L· pk

Then, each section element is defined as:
tt=(2 · al-lla-2) · SIN(te)+2· (1+ lla-al) · (SIN(te))²
bb=(2· al-2· lla-1) · COS(te)+2-(1+lla-al) · (COS(te))³

where:

tt is a vertical element and bb a horizontal one.
lla=tvmax/bvmax
al=(lla/2+1) + (3/4*lla)$^{ggg}$ if lla < 1
al=(lla+1/2) + (3/4)$^{ggg}$ if lla $\geq$ 1
te varies from 0 to ($\pi$ / 2)
     tv = tt· bvmax
     bv = bb· bvmax
ggg is a variable to be chosen to include al in the range of values satisfying the Lewis form, [74, 75].

For fixed values of the main geometric parameters, it is possible to investigate an infinite number of hull form changing either the locations of the four key points or the values of the polynomials powers.

The partial sectional area is then:
av = $\sum$ (2 · tv· bvincr)

and the wetted surface:
ws = $\sum$ (2· tv· incr)

while the mass element is:
mmaa = ro $\sum$ (2· bv· incr· tv· incr)

where incr is the longitudinal step.

a. Calculation of Mass Moments of Inertia. The major Mass Moments of the Inertia of the bare hull are based on the integration of the mass elements mmaa, along the hull.

xlg = L· pk - llvv
xxcc = $\sum$ 2 * bvv * tvincr * iner * xlg
zzce = $\sum$ 2 * bvv * tvincr * iner * tv

where tvincr is the transverse step in each section;
and indicating with vv the volume, the coordinate of the center of gravity are:

xcg = xxcc / vv
zcg = zzcc / vv
and elements of the moments of inertia are:

izz = mmaa · ((2 · bVV)$^2$ + incr $^2$ )/ 12
ixx = mmaa · (incr $^2$ + tv$^2$)/12
iyy = mmaa · ((2 · bvv)$^2$ + tv $^2$ )/12

and, if the center of gravity is not at the origin of the axis relative to the ship, they become:

izc = izz + xlg$^2$ · mmaa - 2 · xlg · mmaa · xcg
ixc = ixx + tv$^2$ · mmaa - 2 · tv · mmaa · zcg
iyc = iyy + (xlg$^2$ + tv$^2$ ) · mmaa 2· xlg · xcg
        mmaa - 2 · tv · zcg · mmaa

Finally, the three moments of inertia are:
$I_z = \sum izc$
$I_x = \sum ixe$
$I_y = \sum iyc$

## 3.3. Conclusions
Based on the results of this appendix, the following conclusions can be drawn:

a) The method contributes a mean for studying effects of hull design changes which may be necessary to improve maneuvering performances of the proposed design well in advance of construction.

b) By coupling the method with a method for calculating the hydrodynamic coefficients, it is possible to study overall performance of various combinations of ship maneuvering designs and systems at the design stage.

d) Once established the main geometric parameters, design changes of the hull can be investigated to improve maneuvering performances.

e) For maneuvering purposes, approximate description of hulls can be represented by polynomials.

## APPENDIX 4

COMPARISON BETWEEN CALCULATION AND MODEL-SCALE MEASUREMENTS OF THE HYDRODYNAMIC SWAY FORCE Y AND MOMENT N EVALUATED BY VARIOUS FACILITIES

### 4.1. Introduction

Maneuvering hydrodynamic forces are usually evaluated either by calculation or by model-scale measurements. This Appendix reports on the comparison of the sway force Y and the yaw moment N between the calculation using the methods given in Appendices 2 and 3 and the model scale measurements of the ship maneuvering hydrodynamic sway force Y and yaw moment N. For this book, two models have been used:

a. ESSO OSAKA, which is a ship for which a number of model test data are · available from other facilities in the world, [87];

b. Series 60, which was a design used in running experimental series and measurements are available from previous studies at the Davidson Laboratory, [88].

The aim is to study the state of agreement/disagreement of the values of hydrodynamics coefficients from calculation and model measurements for deciding if the calculation can be successfully employed in substitution of model-scale testing. Finally, the choice of the coefficients has been constrained by the available data.

### 4.2. ESSO OSAKA

Table 4.1 presents the main parameters of the full scale ESSO OSAKA. Table 4.2 presents a list of facilities that have performed maneuvering tests with a model of the ESSO OSAKA, [87]. Facilities 1 to 11 have used physical models of which the different sizes are reported in Table 4.2, while Facility 12 indicates the calculation based an Appendix 2 and 3 of this book.

Table 4.1
Esso Osaka Full Scale Main Particulars

| | |
|---|---|
| Length overall | 343.00 m |
| Length between perpendiculars | 325.00 m |
| Breadth molded | 53.00 m |
| Draft mean | 21.79 m |
| cb Block coefficient | 0.831 |
| Full load displacement | 328.880 mt |
| Bow bulbous type | yes |
| Stern transom type | yes |
| Number of rudders | one |
| Number of propellers | one |
| Longitudinal CG | 10.30m |

Table 4.2

Facilities and model sizes

| Name | Identification Number | Model size Lpp in m |
|---|---|---|
| Hydronautics | 1 | 7.257 |
| National Maritime Inst. | 2(small model) | 1.625 |
| National Maritime Inst. | 3(large model) | 3.536 |
| Bulgarian Ship Hyd. Cent. | 4 | 8.125 |
| Davidson Laboratory | 5 | 1.625 |
| HSVA-Hamburg/VBD-Duisburg | 6 | 5.000 |
| Sumitoma Heavy Industries | 7 | 2.500 |
| JAMP* | 8 | 3.000 |
| Mitsubishi Heavy Indust. | 9 | 4.600 |
| Nippon Kokan Company | 10 | 6.000 |
| M.I.T. | 11 | unknown |
| Calculation ** | 12 | any |

* JAMP includes: Tokyo University of Mercantile Marine, Hiroshima University, Osaka University
** Method of this book

Table 4.3 presents a comparison of a set of hydrodynamic coefficients of the sway force Y and the yaw moment N. The coefficients are presented in a non-dimensional form and the table illustrates that their values are slightly scattered. The calculated results, facility n. 12, can well be considered as one out of the tank testing facilities.

Table 4.3
Comparison of some principal sway and yaw hydrodynamic coefficients in deep water.

| name of coefficient | value * $10^5$ for facility | | | | | | | |
|---|---|---|---|---|---|---|---|---|
| | 1 | 2 | 3 | 4 | 5 | 6 | 11 | 12 |
| Yv' | -2048 | -1570 | -2030 | -2090 | -2828 | -990 | -2610 | -2790 |
| Yr' | 595 | 1181 | 233 | 816 | 400 | 1870 | 365 | 809 |
| Yv' | -1746 | | | | -1700 | -1694 | -305 | -2032 |
| Yr' | -71 | | | -88 | | 172 | | -146 |
| Nv' | -787 | -696 | -788 | -910 | -1090 | 567 | -1410 | -870 |
| Nr' | -353 | -549 | -216 | -338 | -500 | -219 | -480 | -504 |
| Nv' | -25 | | | -28 | | 321 | | 4 |
| Nr' | -91 | | | -91 | -109 | -37 | | -136 |
| Yvrr' | -1560 | -4139 | -303 | 6700 | -4126 | -2119 | -4500 | -5100 |
| Yvvr' | | 4004 | 1160 | | 1153 | 3634 | | 4260 |
| Nvvr' | | -2098 | -1049 | | -1182 | -3026 | | -4079 |
| Nrrr' | -357 | 894 | 52 | -3000 | 905 | 114 | | 762 |

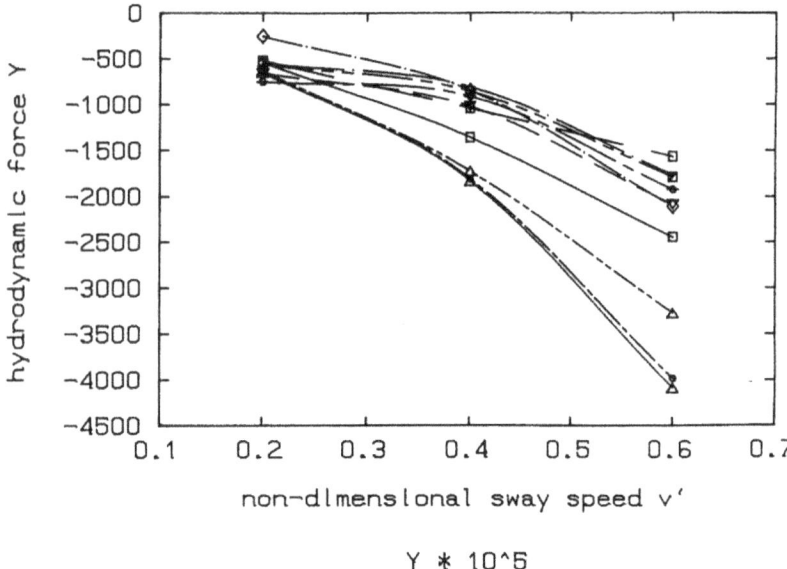

Y * 10^5

Fig. 4.1
Variation of the hydrodynamic force Y with non-dimensional sway speed. Triangles on solid line represent the simulation of this study.

Tables 4.4 and 4.5 present the numerical values of the hydrodynamic force Y and moment N measured and calculated at the different facilities for different values of lateral sway speed and yaw rate. Figures 4.1 and 4.2 are an illustration of the hydrodynamic force. Figure 4.1 illustrates that the Y component is slightly overestimated when varying v' more than 0.25. While Fig. 4.2 illustrates that the N component is averaged over the other results of different testing facilities.

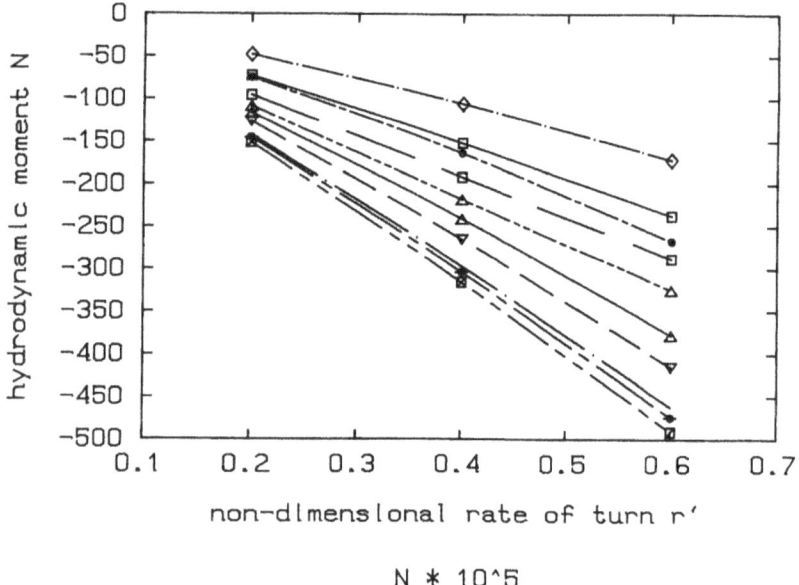

N * 10^5                                        Fig. 4.2

Variation of the hydrodynamic moment N with non-dimensional rate of turn. Triangles on solid line represent the simulation of this study.

Table 4.4
Comparison of deep water sway force and yaw moment

| Facility | $Y'(v')*10^5$ | | | $N'(r')*10^5$ | | |
|---|---|---|---|---|---|---|
| | v'=0.2 | v'=0.4 | v'=0.6 | r'=0.2 | r'=0.4 | r'=0.6 |
| 1 | -545 | -1360 | -2445 | -73 | -152 | -237 |
| 2 | -510 | -1412 | -2706 | -109 | -218 | -325 |
| 3 | -634 | -1726 | -3274 | -45 | -92 | -142 |
| 4 | -478 | -1077 | -1797 | -75 | -164 | -267 |
| 5 | -650 | -1808 | -3982 | -101 | -208 | -327 |
| 6 | -254 | -846 | -2112 | -48 | -106 | -172 |
| 7 | -660 | -1024 | -2092 | -127 | -265 | -415 |
| 8 | -563 | -856 | -1789 | -152 | -316 | -491 |
| 9 | -753 | -913 | -1925 | -146 | -304 | -474 |
| 10 | -550 | -806 | -1764 | -143 | -297 | -462 |
| 11 | -522 | -1044 | -1566 | -96 | -192 | -288 |
| 12 | -648 | -1834 | -4090 | -117 | -241 | -378 |

Table 4.5

Comparison of deep water sway and yaw cross-coupling force

| Facility | $N'(v')*10^5$ | | | $Y'(r')*10^5$ | | |
|---|---|---|---|---|---|---|
| | v'=0.2 | v'=0.4 | v'=0.6 | r'=0.2 | r'=0.4 | r'=0.6 |
| 1 | -157 | -315 | -472 | 128 | 275 | 440 |
| 2 | -178 | -433 | -766 | 219 | 403 | 552 |
| 3 | -173 | -377 | -611 | 45 | 87 | 126 |
| 4 | -164 | -292 | -383 | 170 | 353 | 550 |
| 5 | -215 | -416 | -586 | 84 | 193 | 352 |
| 6 | -22 | -139 | -484 | 375 | 757 | 1154 |
| 7 | -246 | -492 | -738 | -249 | -525 | -827 |
| 8 | -275 | -550 | -825 | -247 | -521 | -822 |
| 9 | -245 | -491 | -736 | -307 | -641 | -1000 |
| 10 | -254 | -508 | -762 | -210 | -446 | -708 |
| 11 | -210 | -420 | -630 | 73 | 146 | 219 |
| 12 | -186 | -444 | -845 | 203 | 416 | 649 |

## 4.3. Series 60

Another comparison of calculation and a model measurements are presented in this section. The Series 60 measurements are from [88]. Table 4.6 presents the main particulars of the model used in this comparison. This comparison's aim is to extend the comparisons to another hull different than the ESSO OSAKA.

Table 4.6
Series 60 Main Particulars

| | | |
|---|---|---|
| Length overall | 1.550 | m |
| Length between perpendiculars | 1.524 | m |
| Breadth molded | .218 | m |
| Draft mean | .813 | m |
| cb Block coefficient | 0.6 | |
| Bow bulbous type | no | |
| Stern transom type | no | |
| Number of rudders | one | |
| Number of propellers | one | |
| Longitudinal CG | .785 | m |

Table 4.7
Comparison of some principal sway and yaw hydrodynamic coefficients in deep water.

| Name of coefficients | Model test | Calculation |
|---|---|---|
| Yv' | 2.185 | 2.018 |
| Yr' | 0.5762 | 0.8003 |
| Yvrr' | -7.9739 | -5.2266 |
| Yrvv' | -4.328 | -2.1186 |
| Yvvv' | -22.934 | -17.148 |
| Yrrr' | 0.689 | 0.851 |
| Yδ' | 0.358 | 0.324 |
| Yδδδ' | 0.3535 | 0.3149 |
| Nv' | 0.6725 | 0.6229 |
| Nr' | -0.4861 | -0.3963 |
| Nvrr' | -0.866 | -1.2241 |
| Nrvv' | -9.915 | -5.226 |
| Nvvv' | -2.143 | -2.6708 |
| Nrrr' | -0.4079 | -0.2880 |
| Nδ' | -0.172 | -0.155 |
| Nδδδ' | -0.1997 | -0.1488 |

The state of agreement/disagreement of each single calculated coefficient can be measured in percentage of the measured one. Generally, for the series 60, such percentage is 30%. While, for Yrvv' and Nrvv', the percentage of disagreement is about 50%.

## 4.4. Conclusions

On the basis of the results of these comparisons the following conclusions can be drawn:

a) The comparisons of the hydrodynamics coefficients from calculation and model measurements demonstrate that their values are slightly scattered.

b) The state of agreement/disagreement between various Esso Osaka models and Series 60 model measurements and calculation demonstrates that the calculated results can be considered as one out of the tank testing facilities.

APPENDIX 5

COMPARISON OF SHIP SIMULATION AND FREE RUNNING MODEL TESTING OF A JACKCLASS TYPE MODEL WITH A DOUBLE VECTWIN RUDDER SYSTEM

## 5.1. Introduction

As a part of this book's work, computer software was developed by extending the work presented in Appendices 2 and 3. The resistance and propulsion contributions are based on statistical evaluation of a large number of model tests, [44, 89-91]. The software was used to run maneuvering simulation tests. Moreover, computer predictions of ship maneuvering performances were used in both case examples of this book to enhance maneuvering-safety characteristics at the design stage. On the other hand, the question of how accurate is the computer prediction at the design stage can only be achieved after comparisons of full-scale measurements of the ship realized from the proposed design. For the scope of this book, free running model testing has been used for:

• investigating the level of agreement and correlation between the computer simulation and physical model measurements;

This appendix describes the comparisons between the computer predictions and physical model measurements of a Jackclass type hull with a sophisticated maneuvering system named Double Vectwin Rudder System. Because the maneuvering system architecture is somehow complicated, in the following the Double Vectwin is described and then the comparison of several trials is provided. Finally, the model test measurements were also aimed to prove the enhanced maneuvering-safety features devised in case example 2.

## 5.2. Description of the Double Vectwin Rudder System

Schilling Rudders, [92, 93], have the ability to control the propeller slipstream. Such features led to the development of twin Schilling Rudders which with conventional propeller, and independent but coordinated movement, [74, 92]. Afterwards, the use of nozzle propeller improved the performance of such a system, [94]. The latest application is the use of a combination of the synchronized vectwin rudder system to achieve total control of the ship movement. This is known as Double Vectwin Rudder System, [94]. By rudder/s movement the degree of the trust from two symmetrical improved nozzle propellers is controlled and the thrust flow is steered through 360 degrees.

The need to reverse shaft direction or propeller pitch is eliminated. Exceptional ship control and maneuverability are obtained with unidirectional engine, shaft and fixed pitch propeller. Each of the rudders is operated by a separate steering gear, 105 degrees outboard and up to 50 degrees inboard. However, the system control is by a single joy-stick which provides the correct angles to control and vector the propeller thrust. In normal navigation the rudders are used in unison controlled usually by a wheel or automatic pilot. Using the unidirectional fixed pitch nozzle propeller, virtually all bridge control for engine and propeller operation is eliminated. Moreover, the system also offers an enhanced stopping ability under full heading control with the engine remaining ahead, stopping distances is at least halved, vibration reduced to a small fraction of that experienced when propellers are reversed.

## 5.3. Description of the Maneuvering Testing of the Proposed Design.

a. Date and Location. On September 3rd and 4th 1992, at the testing tank of the Department of Ship and Marine Technology of Strathclyde University, Glasgow, Scotland, were performed free running model tests of the Jackclass model.

b. Jackclass type model main particulars:

L = 1.56 m
B = 0.53 m
T = 0.14 m
scale 1/30
cp = 0.767

Figures 16 and 17 and photos 1 and 2 in the main text present the line plan of the model. Photos 5.1 to 5.2 are illustrations of the simple but accurate technique used for the test.

c. Joystick. The tests were performed using a remote control joystick. The joystick positions were portions in 360 degrees. Each one representing a particular flow control according to:

Joystick angle             means
= 0 degrees              (max ahead speed)
= 180 degrees           (max astern speed)
= 90 degrees            (max side speed)

other intermediate joystick values have intermediate meanings.

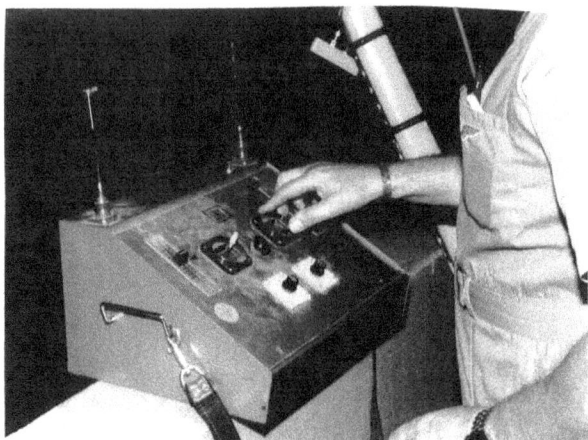

Photo 5.1

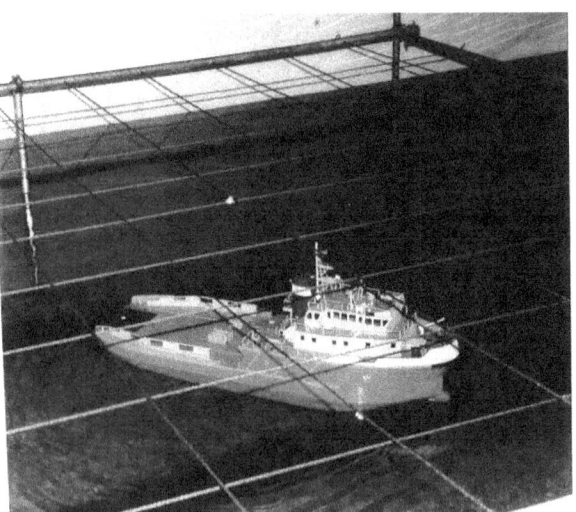

Photo 5.2

d. Type of tests. The following tests were performed:

1. Speed test ahead (Joystick angle = 0)

2. Speed test astern (Joystick angle = 180)

3. Stopping (Joystick angle = 0; RPM =0)

4. Stopping (Joystick angle = 180)

5. Turning (Joystick angle = 45)

6. Turning (Joystick angle = 90)

7. Turning from station-keeping (Joystick angle = 45)

8. Turning from station-keeping (Joystick angle = 90)

9. Sideway motion (two separate joystick controls for
              each couple of rudders)

The model was running at constant RPM on the two ducted propellers (Nozzle type 19, see [93]) using four Schilling rudders.
The model has been operated only changing the rudder angles through the remote joystick control. Test N. 3 was performed with the two propellers set first at full RPM and then at zero RPM.

The use of one joystick made all the maneuvering tests very successful, and besides some radio interferences, they were smooth and not major problems were encountered.

e. Comparison between some free running model testing and simulation. Fig, 5.1 to 5.15 presents the satisfactory agreement between free running model testing and simulation. The stopping test proved very satisfactory. The free running test with Joystick angle at 180 and entry speed of about 1.1 m/s stopped the model in about 5.10 meters in 3.4 seconds equivalent to 3.26 model length. The simulation results of a Jackclass model for a stopping maneuver, with the same setting as the free running model, were about 5.27 meters in 3.51 seconds, equivalent to 3.37 model length. A full scale simulation was also run for the stopping with same setting as the free test but with an entry speed of about 5 m/s. The simulated ship stopped in about 169.65 meters in 60.12 seconds, equivalent to 3.62 ship lengths.

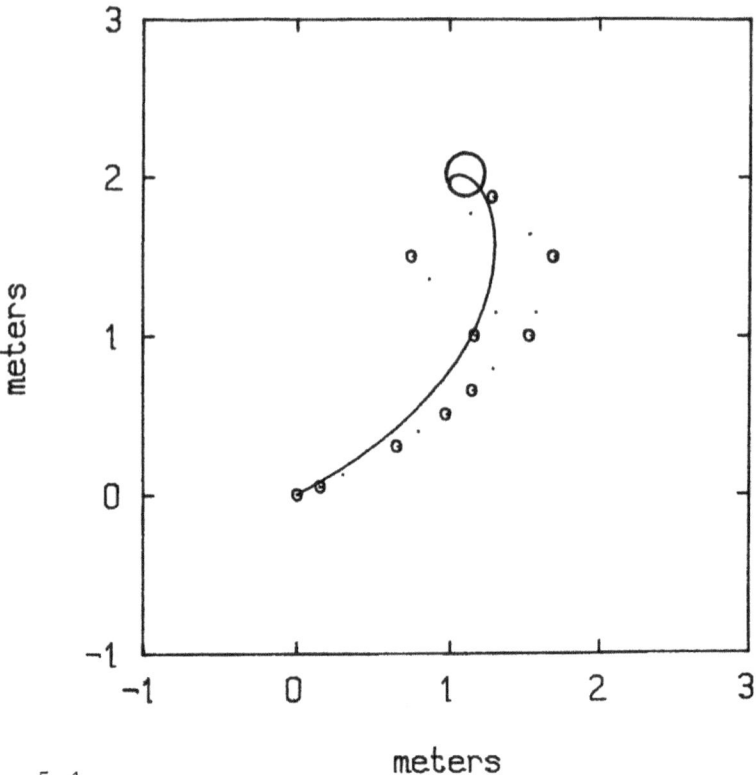

Fig. 5.1

Model trajectory on a turning circle.
Free running model test = circles; simulation = solid line. Approaching speed = 1.59 m/s; Joystick angle = -45.

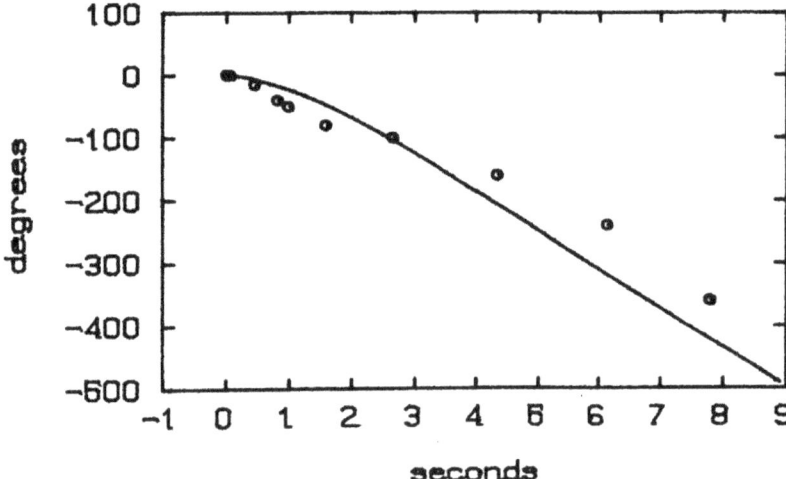

Fig. 5.2
Heading angle during a turning circle.
Free running model test = circles; simulation = solid line. Approaching speed = 1.59 m/s; Joystick angle = -45.

94

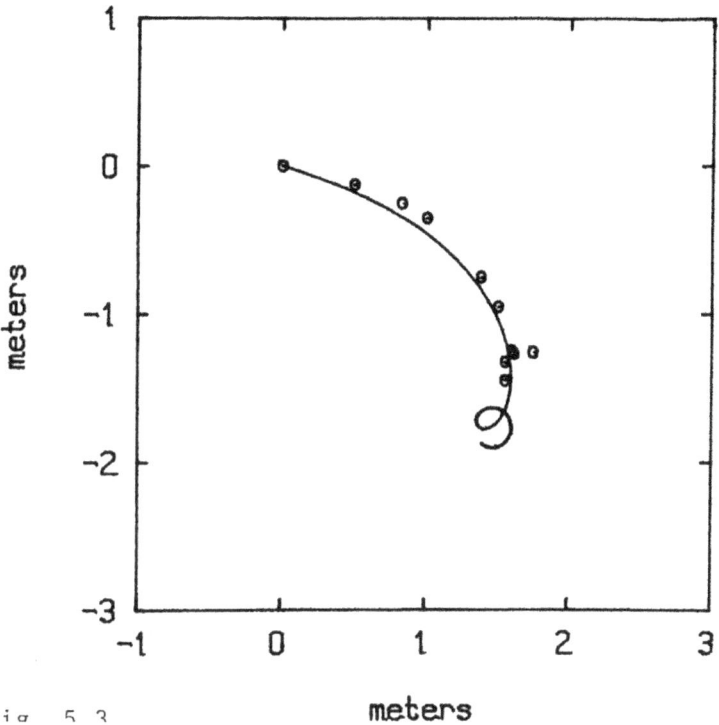

Fig. 5.3

Model trajectory on a turning circle

Free running model test = circles; simulation = solid line. Approaching speed = 1.59 m/s; Joystick angle = +45.

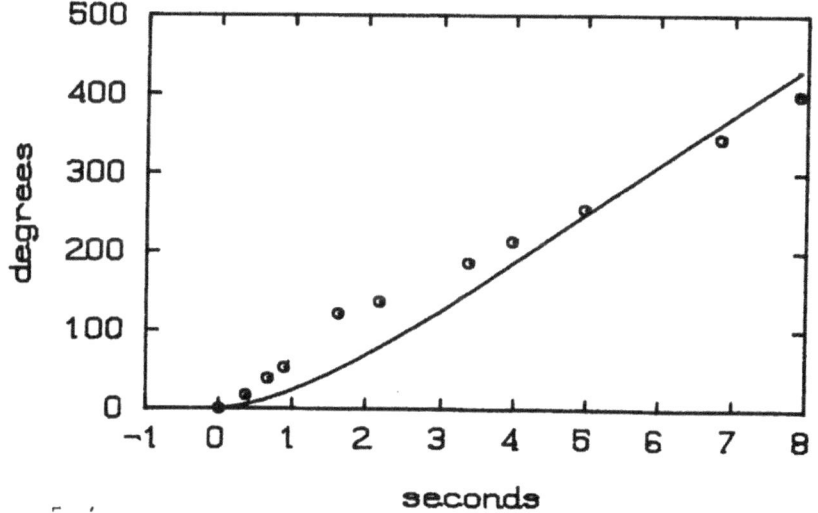

Fig. 5.4

Heading angle during a turning circle

Free running model test = circles; simulation = solid line. Approaching speed = 1.59 m/s; Joystick angle = +45.

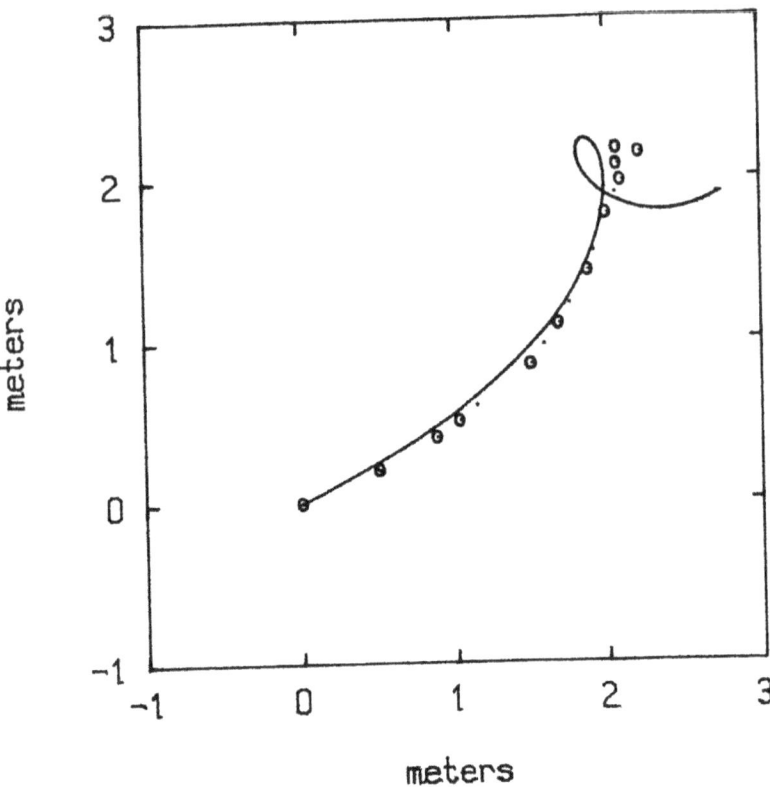

Fig. 5.5

Model trajectory on a turning circle
Free running model test = circles; simulation = solid line. Approaching speed = 1.59 m/s; Joystick angle = -90.

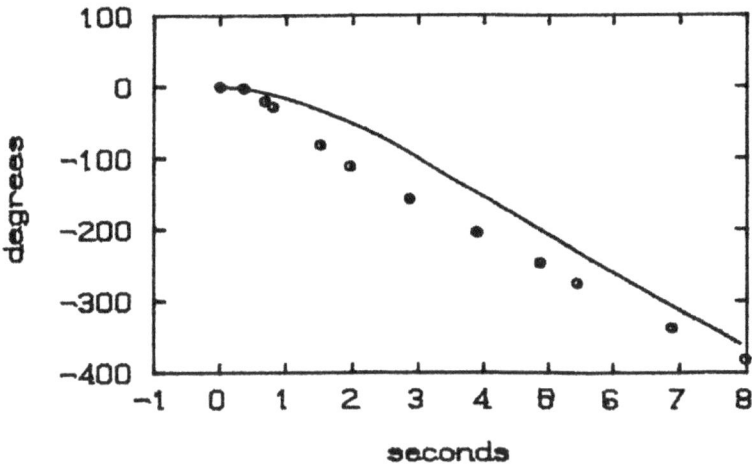

Fig. 5.6

Heading angle during a turning circle
Free running model test = circles; simulation = solid line. Approaching speed = 1.59 m/s; Joystick angle = -90.

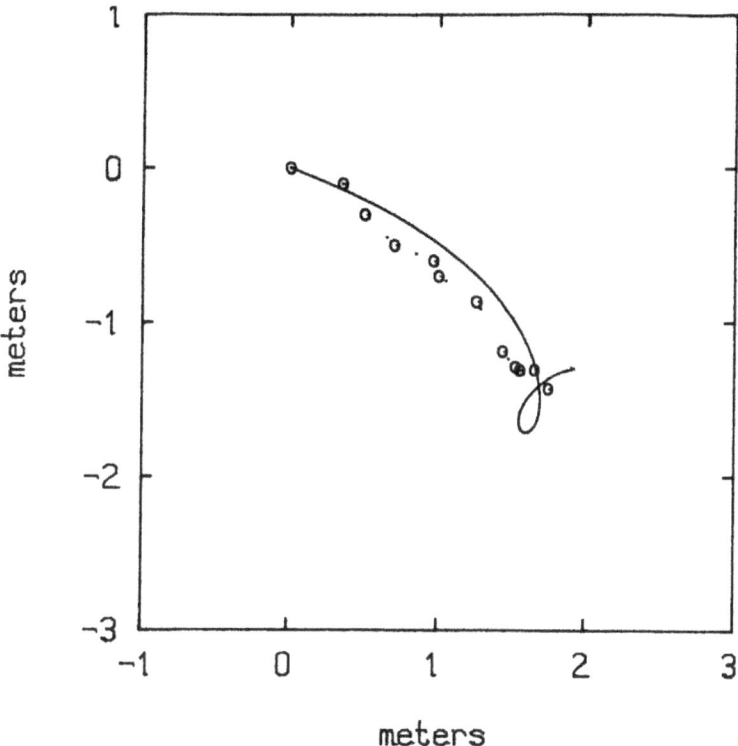

Fig. 5.7

Model trajectory on a turning circle

Free running model test = circles; simulation = solid line. Approaching speed = 1.59 m/s; Joystick angle = +90.

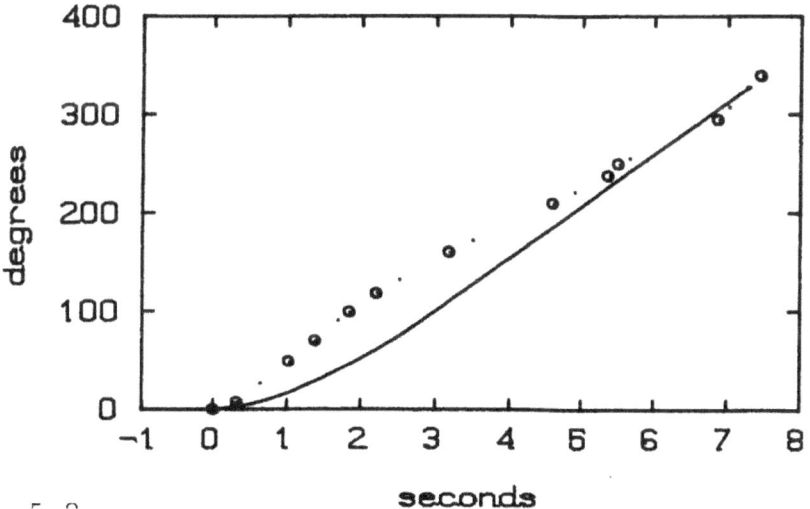

Fig. 5.8

Heading angle during a turning circle

Free running model test = circles; simulation = solid line. Approaching speed = 1.59 m/s; Joystick angle = +90.

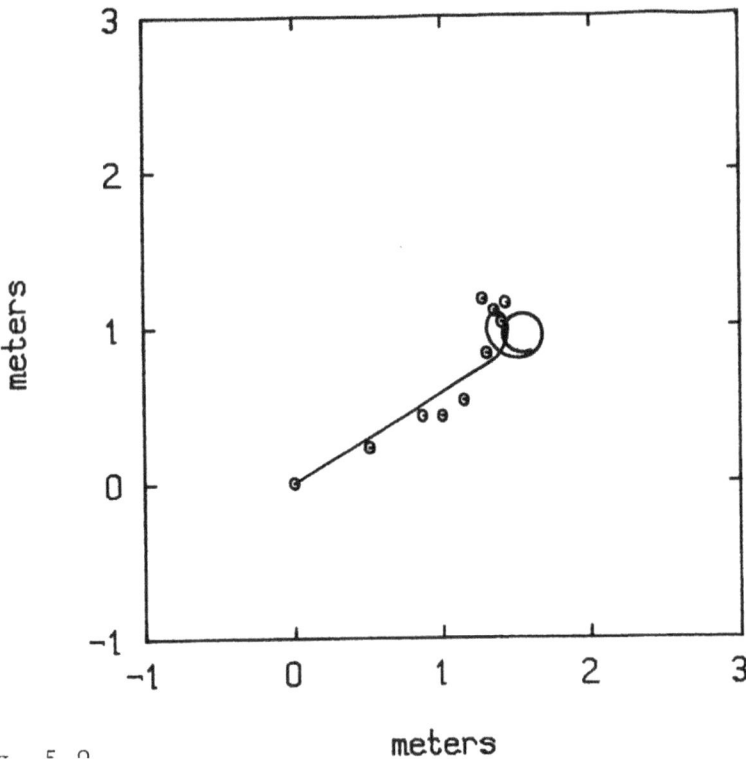

Fig. 5.9

Model trajectory on a turning circle
Free running model test = circles; simulation = solid line. Approaching speed = 0 m/s; initial speed = 1.59 m/s;
Joystick angle = -45.

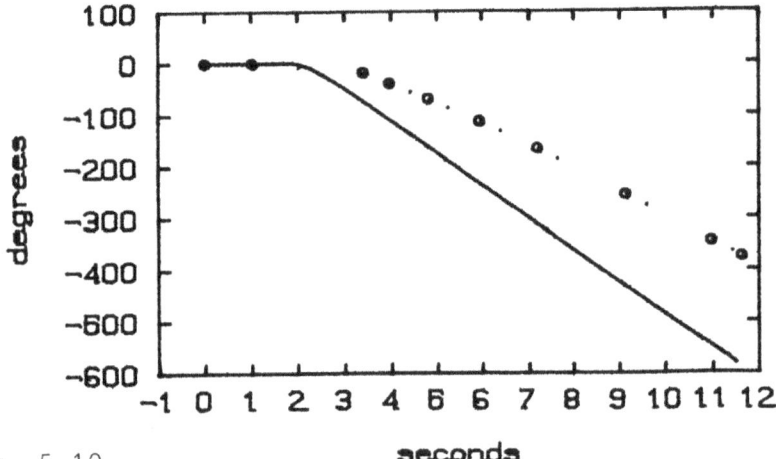

Fig. 5.10

Heading angle during a turning circle
Free running model test = circles; simulation = solid line. Approaching speed = 0 m/s; initial speed= 1.59 m/s;
Joystick angle = -45.

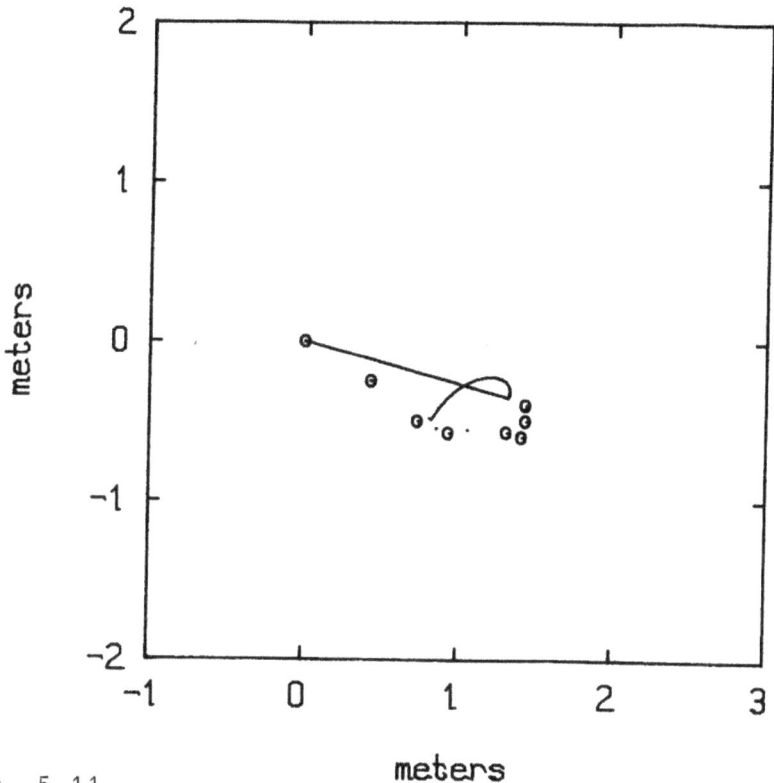

Fig. 5.11

Model trajectory on a turning circle
Free running model test = circles; simulation = solid line. Approaching speed = 0 m/s; initial speed= 1.65 m/s; Joystick angle = -90.

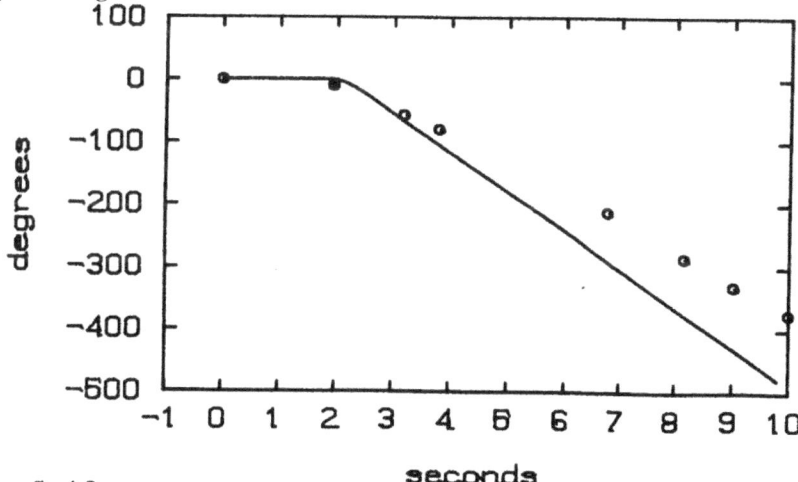

Fig. 5.12

Heading angle during a turning circle
Free running model test = circles; simulation = solid line. Approaching speed = 0 m/s; initial speed= 1.65 m/s; Joystick angle = -90.

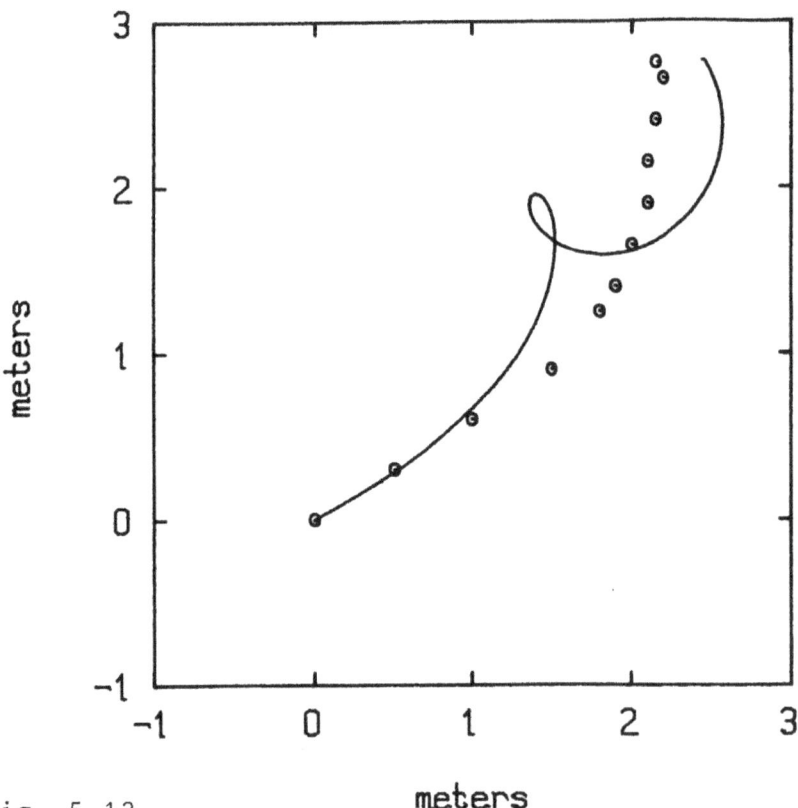

Fig. 5.13
Fig. 5.13
Model trajectory on a turning circle
Free running model test = circles; simulation = solid line. Approaching speed = 1.71 m/s; Joystick angle = -90.

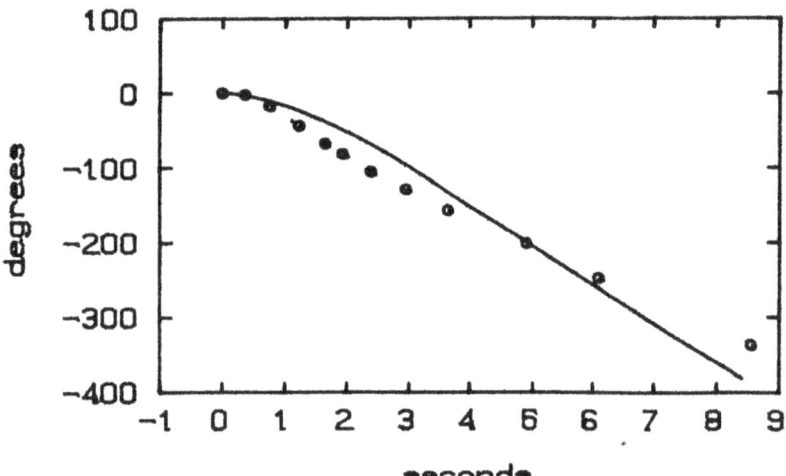

Fig. 5.14
Heading angle during a turning circle
Free running model test = circles; simulation = solid line. Approaching speed = 1.71 m/s; Joystick angle = -90.

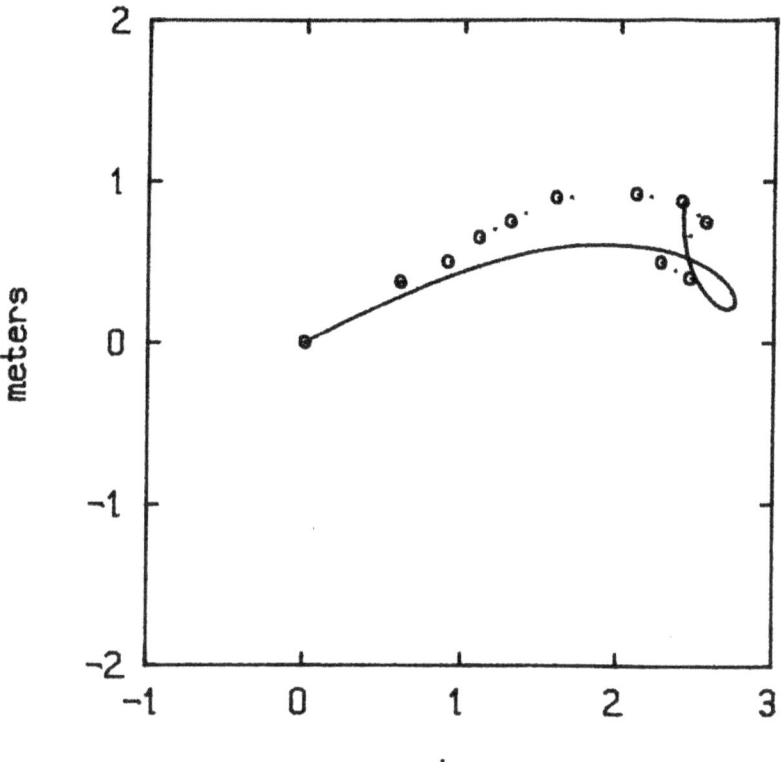

Fig. 5.15                                                    Model trajectory on a turning circle
Free running model test = circles; simulation = solid line. Approaching speed = 1.71 m/s; Joystick angle = 135.

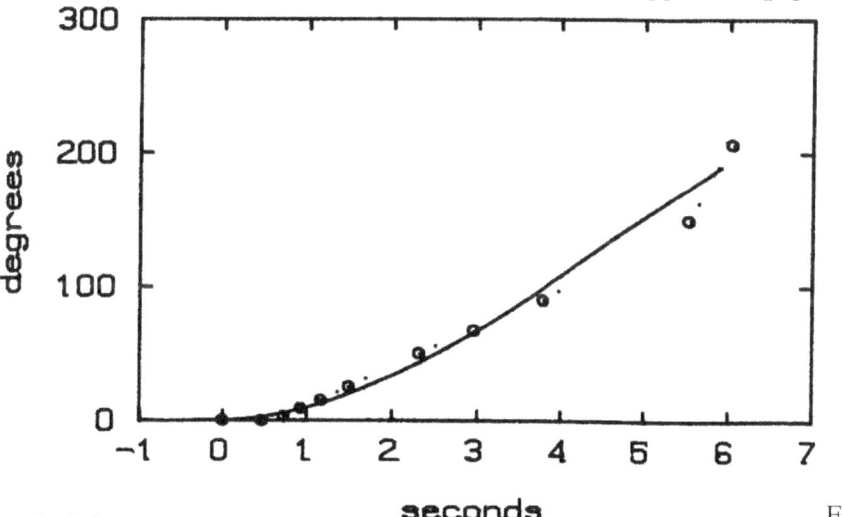

Fig. 5.16

Heading angle during a turning circle
Free running model test = circles; simulation = solid line. Approaching speed = 1.71 m/s; Joystick angle =-135.

## 5.4. **Main Maneuvering Characteristics of the Jackclass Model**

The model shows the following main maneuvering features:

• the model could be stopped at zero speed in about 3 to 5 model length from full speed, (Test N. 4).

• model turning performance from zero speed, after a stopping maneuver, both with Joystick angle at 45 or 90 are very interesting. It has been observed that the model turning motion is very close to a turning point with a slight motion side-way (smaller than the model length), (Tests N. 7 and 8).

• side-way course was observed when operating the model with two joysticks-one for each set of rudders-once the model was stationary the model could move with only sway speed with virtually no ahead or astern speed and not turning motion. However, to keep the model in the proper stationary condition was not an easy task. It requires some training with the two joysticks. (Test N.9).

## 5.5. Conclusions

Based on the results of these comparisons the following conclusions can be drawn:

a) Comparisons between the results of computer prediction and scale-model measurements of a Jackclass type with a sophisticated ship maneuvering configuration demonstrate satisfactory agreement.

b) The good agreement of the comparisons demonstrate that computer predictions of ship maneuvering performances can successfully employed when enhancing maneuvering-safety at the design stage.

APPENDIX 6
THE SHIP'S LIFE CYCLE

## 6.1. Introduction

The purpose of this appendix is to present the phases of a ship's life-cycle. Different sources agree on some phases of a system's life-cycle, [54, 55, 61, 71]. However, the description included in [61] seems more conforming with the management and operation of a ship including the safety aspects.

## 6.2. Ship's Life Cycle

Concept Generation:
The objective of this phase is to prepare a selection of sound possible concepts which would meet the owner's requirements. This will involve tasks ranging from brainstorming sessions through the evaluation of suitable ideas to estimation of approximate costs and the level of safety which can be offered.

Preliminary Design:
The aim here is to establish whether the preferred concept will meet the customer's requirements in the most effective way, so that the owner can decide if the proposed ship is suitable. The alternative name for this phase is "preliminary design". Typical tasks involved are preparing the layout of the ship, calculating estimates of costs involved and considering its safety features.

Detailed design:
The work of this phase should provide all the necessary information for procurement, construction, operation and other relevant purposes. It is sometimes referred to as "engineering" phase and should yield a complete description of the project and its individual components.
This will include the full specification of each part, based on the appropriate code of practice and backed by comprehensive calculations to verify how well the design criteria have been met. The output consists of drawings, material specifications, planning data, procedures, etc. It is also at this stage that details of the resources required are established.

Procurement:
The aim is to purchase from the most suitable sources all the materials and the equipment for the ship, as defined in the previous stage. This involves building up a list of approved suppliers and negotiating with them for the items that are most suitable from the points of view of price and quality.

Construction:
The main aim in this phase is to convert the established details and procured materials into hardware of a ship, with the greatest possible efficiency. The tasks involved include planning, production processes such as cutting, welding and assembling materials into units, the installing of equipment, the control of progress and outfitting.

Commissioning:
The aim is to check that the structure and its equipment will meet the customer's specifications. Acceptance tests and ship trials are performed in this phase and adjustments are made to ensure conditions are satisfied.

Operation:
This is the phase in which the ship has to fulfill the customer's requirements of generating income. Tasks involved include familiarizing the crew with the operation of the ship, close monitoring of performance, maintenance and repairs.

Decommissioning:
After the useful life of the ship is completed the aim here is to dismantle it in environmentally acceptable manner. The tasks involve the systematic down-grading of the functional capabilities of the ship and its eventual breakup.

## 6.3. **Conclusions**

Based on the contents of this appendix the following conclusions can be drawn:

a) The life-cycle of a ship can be represented in 8 phases.

b) Cost and safety aspects have a role to play in each phase, and their maximum impact in the phase of interest will stem from effective decisions made in the earlier phases.

APPENDIX 7
PREVENT-IT SAFETY METHODOLOGY FOR MANOEUVRABILITY

## 7.1. Introduction

The aim of this appendix is to present the steps of PREVENT-IT so that the whole methodology can be readily understood. Moreover, this book requires approaching maneuvering-safety in a way that the results would generate ships which have acceptable maneuvering performances and related with an agreed level of safety.
When integrating maneuvering criteria, the basic ideas of PREVENT-IT can be stated as follows:

"to anticipate inherent or likely maneuvering deficiencies by

(a) preferring ship designs with acceptable maneuvering qualities, and

(b) by taking active steps to minimize the likelihood of their occurrence or to limit their effects;

over the ship's entire life-cycle while satisfying the demand for cost-effectiveness"

## 7.2. PREVENT-IT Steps

The PREVENT-IT consists of nine steps as follows, [26, 60-62]:

In the case of maneuverability, when the PREVENT-IT Safety Methodology is to be applied in each phase of the life cycle, the approach is to select some steps which are very important in each phase. Generally, certain steps may be very important in one phase, while other be more relevant in another. When the decision in one of the step is negative, the steps are repeated until such times a positive decision is reached. Each of the phases of the life cycle will be implemented and the relevant steps of PREVENT-IT applied. To a large extent the steps implemented in a given phase will depend on the amount of information that is available at that time. Following there is a description of the steps of PREVENT-IT for the case of enhancing maneuvering-safety.

Step 1: Predict Potential Maneuvering Hazards
The first step is to predict potential maneuvering hazards for the concept, design or ship as appropriate. Techniques such HAZOP and HAZAN are applicable, [95, 96]. It should be noted that the range and extent of potential maneuvering hazards which can be identified will depend on which phase of the life cycle is under examination. Typical maneuvering hazards are collisions and contacts.

Step 2: Research into Maneuvering Risks
The maneuvering risk of occurrence should be calculated for each of the hazards identified.
Consequences of maneuvering failures are determined by a combination of the following:

• understanding the hydrodynamics of the ship in question;

• applying maneuvering experience gained from similar ships;

• benefitting from simulation of the ship's maneuvering behavior.

Probability of occurrence is determined by a combination of statistics and experience. Risk is defined as the product of consequence by probability. Typical techniques such as Fault Tree Analysis (FTA) can be used, [48, 49].

Step 3: Establish the Role of Human Factors in Maneuvering Operations

This is a critical step in the methodology because it is concerned with human behavior. The human factors of interest in the present context are:

• management features: these are related to matters such as policies, maneuvering-safety culture and the communication mechanism within the organization;

• competence features: these are related to the skills of individual employees and involve such matters as the ability to understand fundamental principles, acquire knowledge, apply information and solve problems;

• operational features: these are related to activities associated with the maneuvering operation of the ship and involve features such as maneuvering procedures, maneuvering criteria, the feedback of information and experience.

All three sets of features must be considered from the point of view of maneuvering-safety.

It should, however, be borne in mind that a high standard of maneuvering-safety can only be achieved by effective provision for the human factors combined with the engineering of the ship.

Step 4: Verify the Scope for Design Modifications
On the basis of the previous three steps it may be possible to reduce maneuvering risks by careful design. Changes are best introduced at an early stage of the design stage – say the first two phases of the ship's life cycle.
The aim of the changes will head for greater simplicity reducing the number of maneuvering features likely to go wrong, while improving the ease of maneuvering operation and minimizing the cost of maintenance.

Step 5: Engineer the Maneuvering Containment System
Research has shown that many factors contribute to the occurrence of a major accident, [1-6].
Various maneuvering factors are called as main causes in several cases, [3-6]. Moreover, Charlton, [47], has shown how thirty interacting factors were involved in one case.
Clearly it would be impossible to cater for every eventuality but introducing suitable containment solutions or preventing certain of the crucial factors from being activated simultaneously, it may well be sufficient to reduce a potential major accident to a minor one, and to change a potentially minor accident to a "non-accident". Particular rudders or special type of propellers may be fitted to reduce specific crucial factors such as turning diameters or propeller efficiency at slow speed.

Step 6: Nominate Viable Maneuvering Emergency Solutions
Emergency solutions will always be required, no matter how far potential maneuvering hazards are minimized. Typical solutions in ships include the use of underwater braking devices to reduce stopping distances. Supplementary rudders' engine can be made ready to act in case of the rudder's engine breakdown during maneuvering operations.

Step 7: Transmit Quality Requirements
Failures often occur because design details were not correctly fulfilled or maneuvering procedures were not followed, i.e. quality implementation is not achieved. "Quality" is defined here as "the ability to meet the agreed maneuvering specifications". A ship designed for a specific maneuverability, cannot be expected to operate satisfactory in other modes. For example, it is not possible to operate a large tanker similarly to a supply vessel. In practice, quality performance is required in every aspect of an organization's operations but the following factors are most important:
• a positive attitude to quality;
• understanding what quality operation involves;
• having the techniques to implement quality procedures and measures
performance;
• providing feedback;

Although these requirements have always existed much greater attention is being given to quality today through the introduction of the concept often called Total Quality Management, [97] .

Step 8: Interface with Maneuvering and other Regulations The previous steps can be regarded as the "good practice" of any organization. But ships have to satisfy a variety of national and international rules and regulations as well, [56-

59]. It is essential that these maneuvering requirements are satisfied, although in many practical situations the maneuvering-safety standards of the organizations involved may be higher than the statutory requirements.

Step 9: Train the Personnel

To ensure that a high standard of maneuvering-safety is maintained, all personnel must be trained to recognize potentially maneuvering dangerous situations, to take appropriate action to correct them, and to provide feedback on experience gained in practical situations. Use of "full-mission" maneuvering simulator can be very effective to achieve high level of maneuvering training for seagoing personnel.

## 7.3. Conclusions

Based on the contents of this appendix the following conclusions can be drawn:

a) A higher standard of maneuvering safety can be achieved through a balanced combination of features relating to management, engineering and operation.

b) Many aspects and activities enforced by the PREVENT-IT could already be part of good design practice, however it is better to have a method for implementing it.

APPENDIX 8
DESIGN CHANGES OF CASE EXAMPLE 1

## 8.1. Introduction

This appendix presents the methodology employed to implement the design changes of case example 1 of the book. The previous studies given in [98, 99] have demonstrated a method of enhancing maneuverability by studying the variation of hull forms and appendages with calculation techniques. The results of these studies have been given in [8], and it appears that they have obtained reasonable agreement with model scale results.

In this book, to enhance maneuvering-safety, the proposed design changes of the preliminary design phase of the example requires that the sway speed v should decays faster to zero than the yaw rate r. In such a way that the sway speed is attenuated more rapidly than the yaw rate and the hull turns onto the new course with a smaller drift. In Chapter 7, it has been discussed that the magnitude of the Q-index is smaller for ships that maintain lateral velocity than it is for ships whose lateral velocity varies during a turn. A large magnitude of Q indicates that the vessel will sway significantly in response to even small changes in r. The objective of the design changes is to devise a preliminary design with improved values of the Q-indices as smaller as possible. In order to achieve enhance maneuvering capability, the following procedure is used:

a) Design Changes of the Hull:
Identify and devise the optimum bare-hull forms;

b) Design Changes of the Active Controls:
Identify and devise the optimum type and location of rudder and location of the propeller.

## 8.2. Design Changes of the Hull

The strategy to be used is:

1. fix geometric parameters
2. select bow design
3. select stern design

The chosen hull's parameters satisfy the maneuvering safety criteria. Indeed, the result satisfies the rules of Appendix 1:

$W_H$= -5.12 < 5.32.

Thus, the point (cb, B/L) = (0.58, 0.144) lies below the classification line in Figure 8.1.

On the other hand, the bow and the stern are complex shapes. To select the optimum form, it is necessary to evaluate the maneuvering indices. This calculation requires hydrodynamic coefficients that take into account complicated geometric relationships. A technique developed for this study is presented in the Appendix 2, and it is employed to carry out these calculations. Remarkably good agreement has been found between coefficients computed by this method and coefficients computed by model testing, see Appendix 4.

To simplify the example, one standard bow design has been adopted, see Figure 8.2. In addition, a limited number of stern designs are further investigated. The LCB (Longitudinal Center of Buoyancy) and Moments of Inertia with respect to the three major ship axes are given in Table 8.1. The five hull forms are the same for the first 120 meters measured from the fore-point of the bow.
While, five different stern forms are used for the remaining 41 meters. The five different stern forms are displayed graphically in Figures 8.3 to 8.7. In the figures, each transverse section corresponds to a distance of 5 meters starting from the fore point of the bow that is considered = 0.

The values of $Q_H'$ for the five designs are given in Table 8.2. On the basis of these values, hull form number 1 is selected for further investigation.

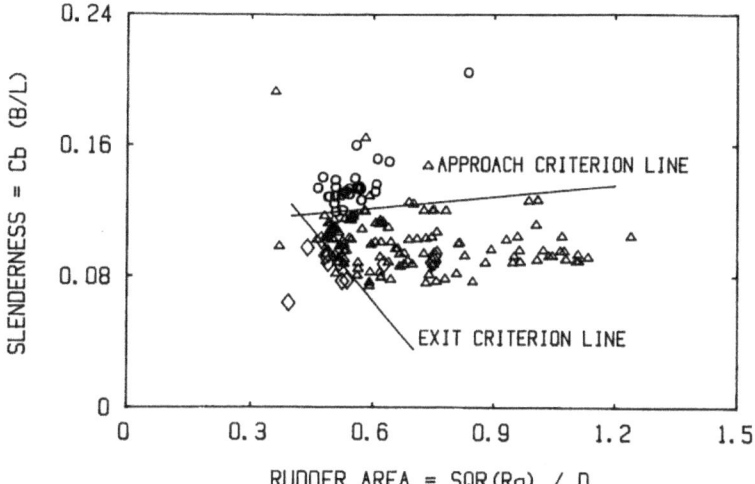

Fig. 8.1
Each point represents a ship in the database of 173 ships. Triangles represent ships whose hull-rudder and rudder indices are normal. Ships that exhibit abnormal or unacceptable behavior are represented by circles and diamonds. Circles correspond to ships whose hull-rudder index is abnormal. Diamonds correspond to ships whose rudder index is abnormal. See Chapter 7 and Appendix 1.

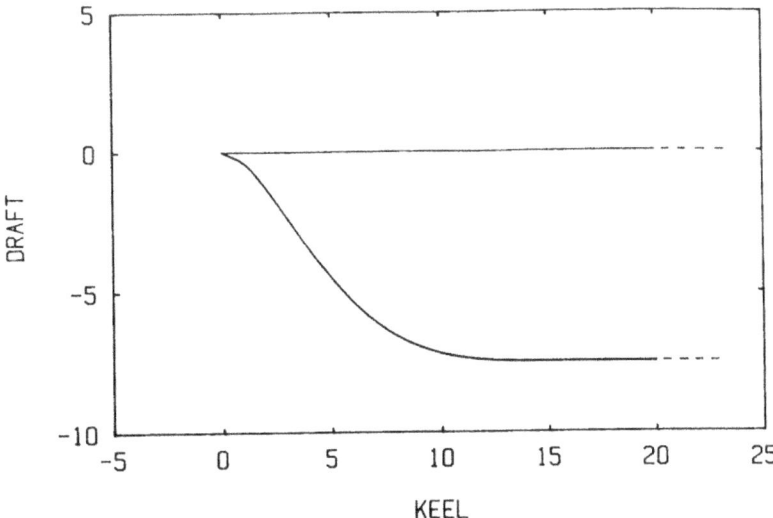

Fig. 8.2

Adopted bow design.

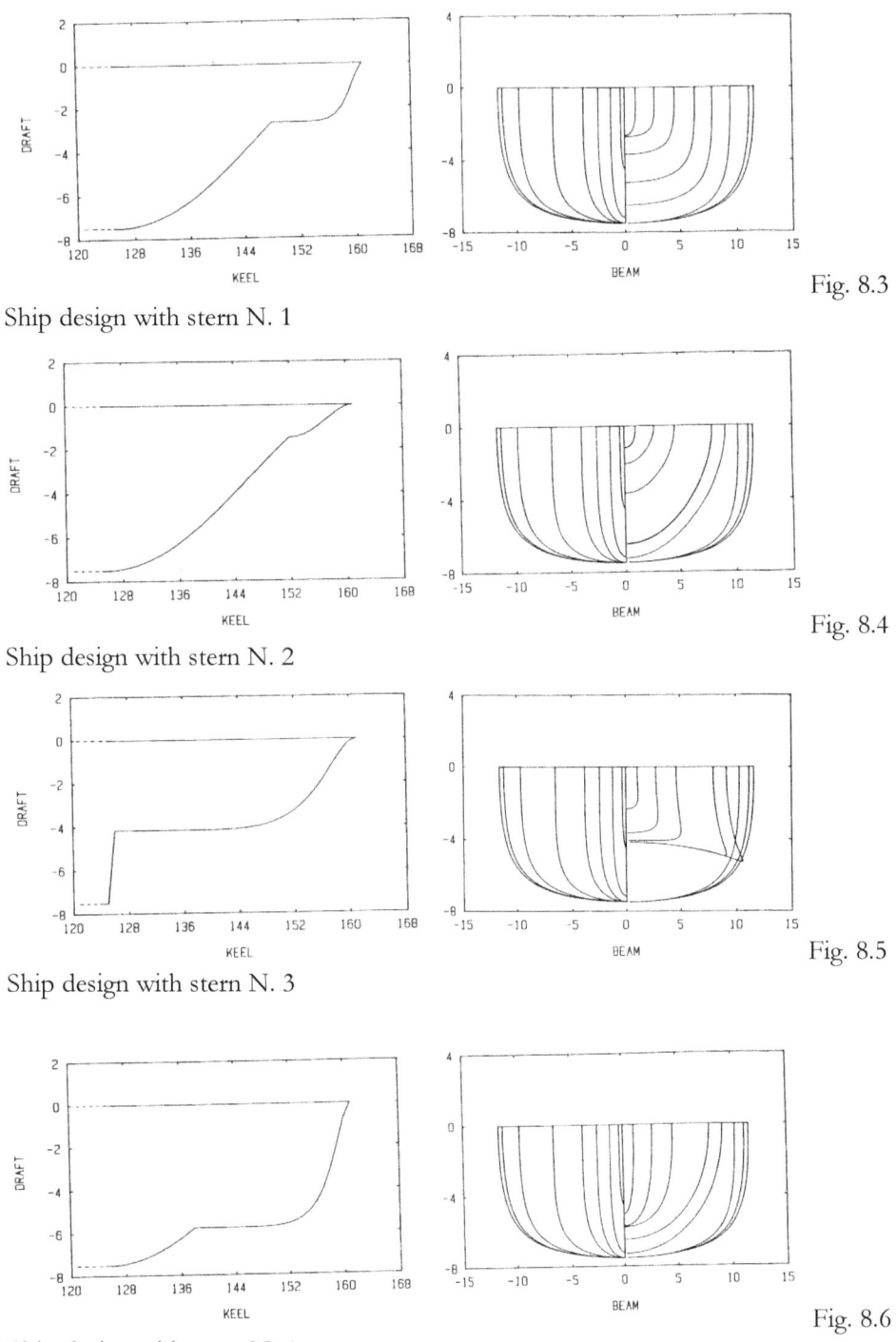

Fig. 8.3

Ship design with stern N. 1

Fig. 8.4

Ship design with stern N. 2

Fig. 8.5

Ship design with stern N. 3

Fig. 8.6

Ship design with stern N. 4

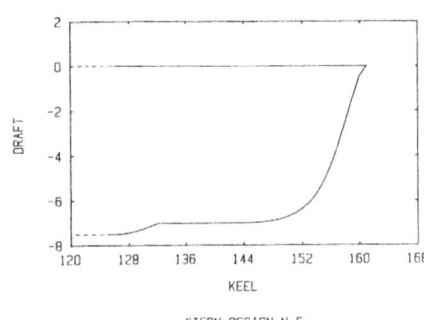 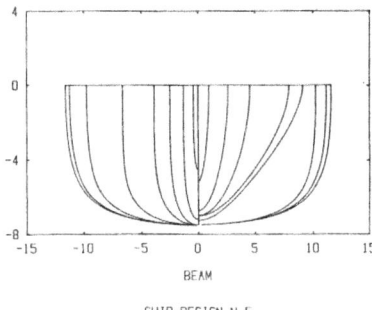

Fig. 8.7

Ship design with stern N. 5

Table 8.1

The LCB and the Moments of Inertia with respect to the three major ship axis of the proposed design with 5 different sterns

| Stern N. | LCB | $Iz'*10^{-4}$ | $Ix'*10^{-7}$ | $Iz'*10^{-4}$ |
|---|---|---|---|---|
| 1 | -.0178 | 3.40 | 6.66 | 3.41 |
| 2 | -.0174 | 3.40 | 6.67 | 3.42 |
| 3 | -.0472 | 3.85 | 4.18 | 3.85 |
| 4 | -.0281 | 3.54 | 6.41 | 3.55 |
| 5 | -.0319 | 3.54 | 6.79 | 3.54 |

Table 8.2.

Bare hull steady state index for the 5 different stern designs

| Ship Design N. | 1 | 2 | 3 | 4 | 5 |
|---|---|---|---|---|---|
| $Q_H'$ - Index | -.290 | -.298 | -.430 | -.370 | -.445 |

## 8.3. **Design Changes of the Active Controls**.

The first step in selecting the rudder is to fix the rudder aspect ratio $\sqrt{Ra}/D$. The variation of $Q_{HR}'$ and $Q_R'$, see Chapter 7 of the book, with respect to the aspect ratio are shown in Figure 8.8 for the ship under examination. Inspection of the figure indicates that the $Q_R'$ (RIND on the figure) is insensitive to variation of the ratio, and that QHR' decreases as the ratio increase. Consequently, the rudder aspect ratio should be set as small as possible. Indeed, the point $(S_1, \sqrt{Ra}/D)$ has to lie above the exit criterion of Fig. 8.1. The values of L, B, and cb, indicating that $S_1$ = .084, and the W-rule $W_{HR}$=-2.493-2.951($\sqrt{Ra}/D$), see Appendix 1, lead to the choice of:

$\sqrt{Ra}/D = .535$

Since the value of the draft D has already been set, see Table 8.1, the value of the rudder area will be Ra = 16.1 m².

Rudders of differing design can have the same area. For instance, the area Ra of each of the rectangular rudders shown in Figure 8.9 is 16.1 m².

Variations of cr and br generate different rudder designs. The second step in the selection of the rudder is to fix the rudder chords and effective span. Variation of the maneuvering indices with respect to these variables is shown in Figures 8.10 and 8.11. Increasing the rudder chord (cr) for the fixed Ra improves the value of $Q_{HR}$ (Q-INDEX on the

111

figure). Increasing the rudder span improves $Q_R$ (R-INDEX on the figure). The values of the maneuvering indices corresponding to the rudder configurations shown in Figure 8.9 are tabulated in Table 8.3.

Rudder type 1 is selected for its better performances, it has the largest value of $Q_{HR}$ but the smallest value of $Q_R$. The final step in designing the rudder is to increase $Q_R$ by placing the rudder in an optimum location.

Variation of the maneuvering indices with rudder location is shown in Figure 8.12. Inspection of the figure indicates that $Q_R$ can be improved by reducing the value of $xr/L$. the values is set at -0.472.

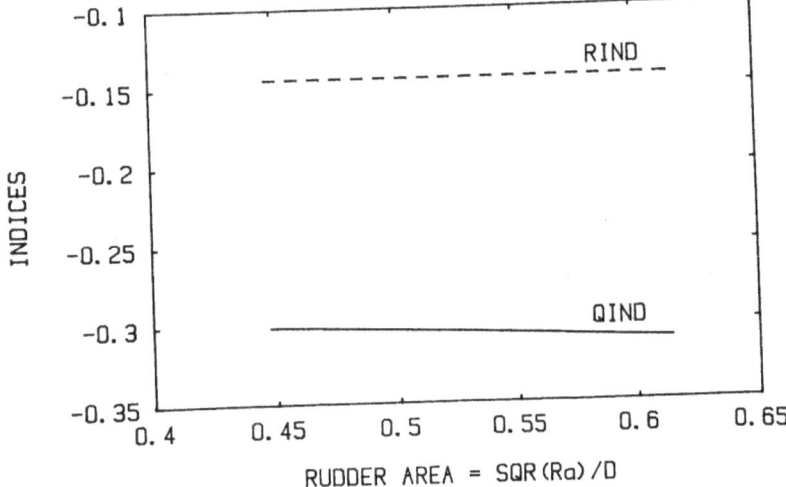

Fig. 8.8

Bare hull (stern N. 1) with rudder. Variation of $Q_{HR}'$(QIND) and $Q_R'$(RIND) with rudder aspect ratio.

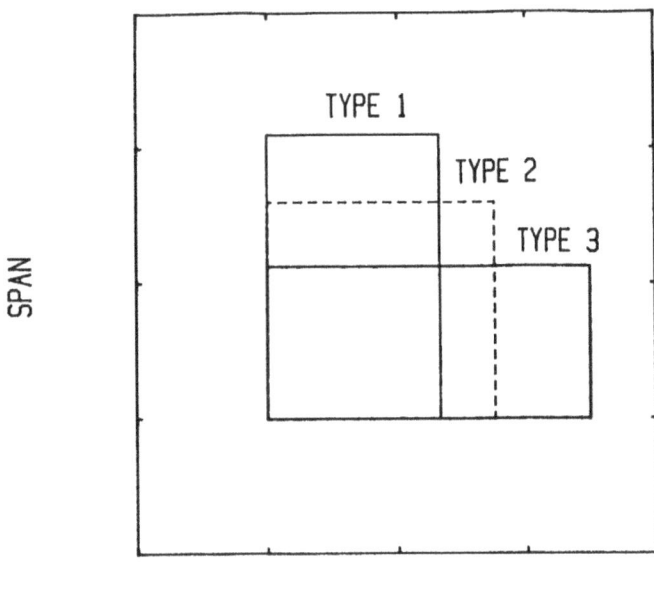

Fig. 8.9

Three rectangular rudders with the same area (16.1 m²)

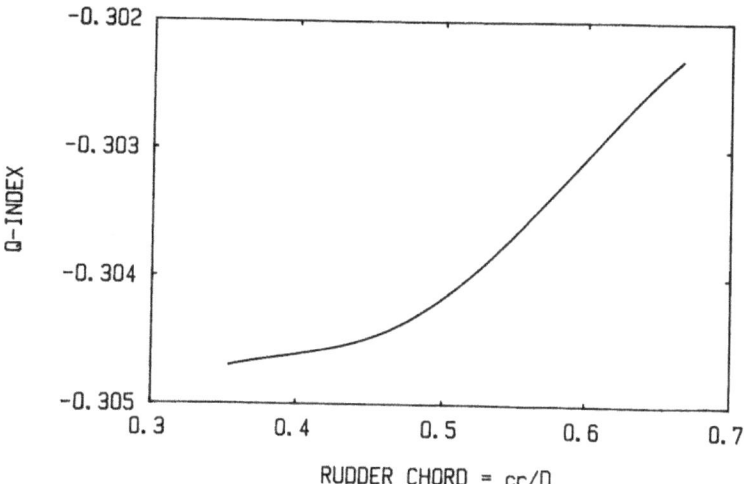

Fig. 8.10

Bare hull (stern N.1) with rudder of area 16.1 m²
Variation of $Q_{HR}'$(QINDEX) for different rudder chords (cr).

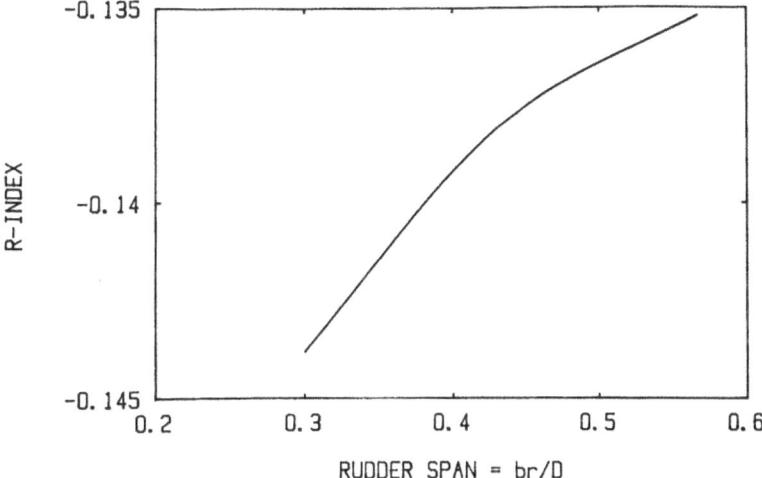

Fig. 8.11

Bare hull (stern N.1) with rudder of area 16.1 m².
Variation of $Q_R'$(RINDEX) for different rudder spans (br).

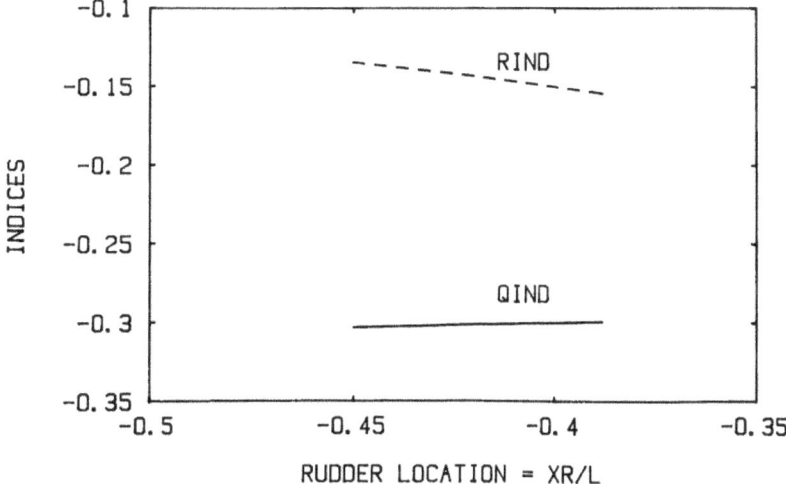

Fig. 8.12

113

Variation of the maneuvering indexes with rudder location for a rudder type N. 1.

Table 8.3
Maneuvering indexes for the three rudder types.

| Rudder Type | 1 | 2 | 3 |
|---|---|---|---|
| $Q_{HR}'$ | -.303 | -.305 | -.307 |
| $Q_R'$ | -.143 | -.135 | -.138 |

The type of propeller selected will influence the ship's ability to turn. It is necessary to keep the maneuvering indices within acceptable bounds. Figure 8.13 depicts the variation of $Q_{HR}$ with propeller diameter. Figure 8.14 shows the dependence of $Q_{HR}$ on propeller location. $Q_R$ is not affected by the propeller diameter. For the hull-rudder configuration under examination results that the index is relatively insensitive to the propeller diameter and location. So it can be chosen based on geometric constrains imposed by the rudder and stern design.

## 8.4. Results

This appendix has described the design changes of case example 1. In the following Table 8.4 the main parameters of the final design are presented.

Table 8.4.
Main parameters of the proposed ship design

```
Ship Type N.   1
Rudder Type N.  1
√Ra/D=.535
xr/L=-.472
xp/L=-.421
dp/T=.466
QH'=-.290
QR'=-.130
QHR'=-.307
```

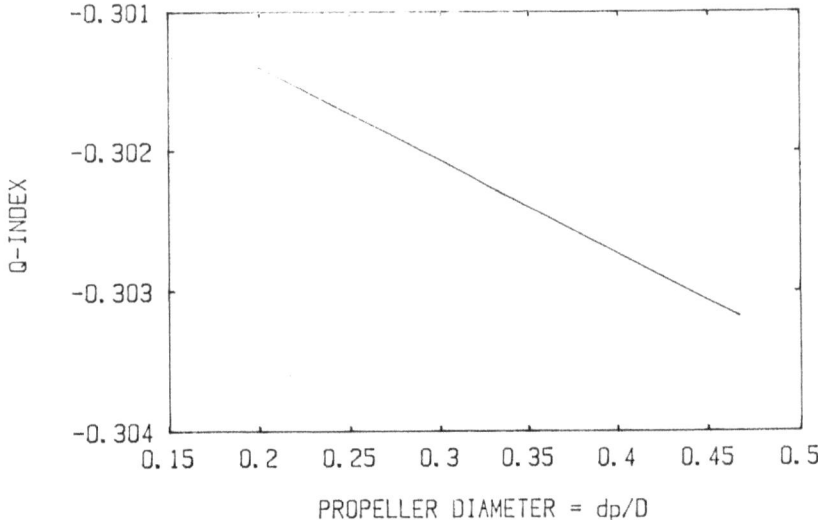

PROPELLER DIAMETER = dp/D

Fig. 8.13

Bare hull (stern N.1) with rudder and propeller.
Variation of $Q_{HR}'$ for different propeller diameters.

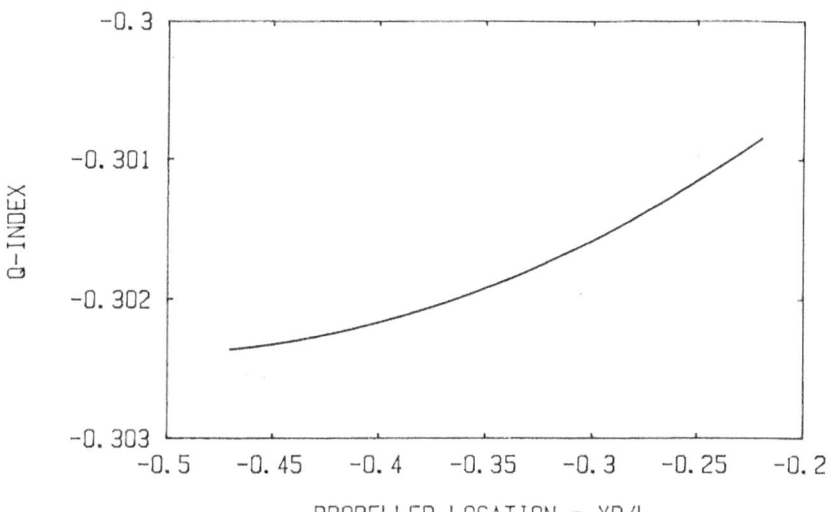

Fig. 8.14

Bare hull (stern N.1) with rudder and propeller.
Variation of $Q_{HR}'$ for different propeller location.

APPENDIX 9
CALCULATION OF RISK FOR CASE EXAMPLE 1

The calculation of the risk value (R) for the Risk Analysis of STEP 2 of the Concept Generation and the Preliminary design phases of Case Example 1 are described in this Appendix. For this Risk Analysis, it is assumed that

Risk (R) = Probability of Occurrence (P) x
                       Consequences Probability (C)

## 9.1. Calculation for STEP 2 of the Concept Generation

Probability of Occurrence. Published statistics, [76], reports that 18.91 per 1000 ship-years of dry cargo ships are involved in maneuvering hazards with the following subdivision:

Hazard type
1. Grounding        10.19 x 1000 ship-years
2. Collision        6.64 x 1000 ship-years
3. Contact          2.08 x 1000 ship-years

Reports [51, 77] show the distribution of the probability of occurrence on the ship length of dry cargo ships.

Particularly, for the three concepts under examination the following percentages are reported:
Concept A: L=161 m        10.6%
Concept B: L= 70 m        27.3%
Concept C: L=100 m        33.3%

Combining these percentages with the previous statistics for hazard type, the following probability of occurrence is obtained:

Table 9.1
Probability of Occurrence
Per 1000 ship-years

| Hazard Type | Concept | | |
|---|---|---|---|
| | A | B | C |
| 1 | 1.08 | 2.78 | 3.39 |
| 2 | 0.70 | 1.81 | 2.21 |
| 3 | 0.22 | 0.57 | 0.69 |

If the probability of occurrence is expressed in the scale 0 to 1 as follows:

1= 3.39 per 1000 ship-years
0=0 "                              "

Table 9.1 becomes as follows:

Table 9.2
Probability of Occurrence

| Hazard Type | Concept | | |
|:-----------:|:------:|:------:|:------:|
|             | A      | B      | C      |
| 1           | 0.31   | 0.82   | 1      |
| 2           | 0.21   | 0.53   | 0.65   |
| 3           | 0.06   | 0.17   | 0.13   |

Consequences. Consequences can be classified using a "Severity Classification" to provide a qualitative indicator of the worst potential effect resulting from the hazard. Applying the results described in [53], it is possible to extent the severity of consequences to ships as follows:

Table 9.3
Severity Classification

| Category | Name | Consequences |
|:--------:|:-----|:-------------|
| I | Catastrophic | Loss of Ship, Death |
| II | Critical | Major Ship Degradation, severe injury, damage or reduction in mission performance |
| III | Marginal | Minor Ship Degradation, minor injury, minor damage or reduction in mission performance |
| IV | Minor | No injury or ship degradation, but may result in ship failure and unscheduled maintenance or repair |

However in this calculation the consequences probability C is used. The value C represents the conditional probability that the consequences with the specified critically classification of Table 9.3 will occur given that the hazard occurs. Usually for complex systems such as ships, C is difficult to calculate and thus becomes a matter of judgment, meaning it is greatly driven by the analyst's prior experiences. However, C can be based on the severity of the hazard as shown in Table 9.4, [53]:

Table 9.4

| | Linear Systems | Non-linear Systems |
|---|---|---|
| Severity | C | C |
| I | 1 | 1 |
| II | 0.75 | $0.1 < C \leq 1$ |
| III | 0.5 | $0.05 < C \leq 0.1$ |
| IV | 0.25 | $0 < C \leq 0.05$ |

Published statistics, [51, 75, 77], reports that for each hazard type the following percentage of serious casualties and total loss are reported:

| Hazard | Serious Casualties | Total Loss |
|---|---|---|
| 1 | 72.4% | 27.6% |
| 2 | 87.7% | 12.3% |
| 3 | 89.9% | 10.1% |

Where the distribution by ship length gives the following percentage for each concept:

| Concept | Serious Casualties | Total Loss |
|---|---|---|
| A | 100% | 0% |
| B | 95% | 5% |
| C | 70% | 30% |

Table 9.5 reports the value of C in terms of the relationships of Table 9.4.

Table 9.5
Consequences Probability

| Hazard | Concept A | B | C |
|---|---|---|---|
| 1 | 0.65 | 0.75 | 0.9 |
| 2 | 0.6 | 0.7 | 0.85 |
| 3 | 0.55 | 0.65 | 0.8 |

9.2. **Calculation for STEP 2 of the Preliminary Design**

Probability of Occurrence. Published statistics, [51, 76, 77], reports that 2.21 per 1000 ship-years of dry cargo ships are involved in collisions and contacts with the following percentage subdivision:
Hazard type:
Collisions and Contacts
1: in open waters      21%
2: in restricted waters 43%
3: in port               36%

Reports [51, 77] show the distribution of the probability of occurrence on the ship length of dry cargo ships. Particularly for concept A under examination the percentage of 10.6% is reported, i.e. 0.234 per 1000 ship-years, so that combining these percentages with the hazard type, the following probability of occurrence is obtained:

Table 9.6
Probability of Occurrence
Per 1000 ship-years

| Hazard Type | Concept A |
|---|---|
| 1 | 0.049 |
| 2 | 0.100 |
| 3 | 0.084 |

and using the same class of probability of occurrence as the previous section 9.1.:

Table 9.7

| Hazard Type | Concept A |
|---|---|
| 1 | 0.014 |
| 2 | 0.029 |
| 3 | 0.025 |

Consequences. Published statistics, [51, 75, 77], reports that for each hazard type the following percentage of serious casualties and total loss are reported:

| Hazard | Serious Casualties | Total Loss |
|--------|--------------------|-----------|
| 1 | 87.5% | 12.5% |
| 2 | 74.5% | 25.5% |
| 3 | 85.7% | 14.3% |

In Table 9.8, the calculated values of C in terms of the relationships of Table 9.4 are reported.

Table 9.8
Consequences Probability

| Hazard Type | Concept A |
|-------------|-----------|
| 1 | 0.6 |
| 2 | 0.8 |
| 3 | 0.7 |

APPENDIX 10
DEFINITIONS OF SYMBOLS

## Hull

Ab  geometric base area of hull
Abe  effective base area hull
Abf  lateral projected area of bow fin
Abi  lateral area of the bow of body of revolution
Af  frontal area of the hull
Atf  lateral projected area of tail fin
Ati  lateral area of the tail of body of revolution
Ax  frontal area of submerged portion of hull
ARtf  aspect ratio of tail fin
ARbf  aspect ratio of bow fin
B  hull beam at waterline
cp  prismatic coefficient
cb  block coefficient
cm  midsection coefficient
db  diameter of the base of body of revolution
dbe  effective diameter of base of body of revolution
hd  T
hbf  bow draft
htf  tail draft
$I_z$  mass moment of inertia with respect to vertical axis
$I_x$  mass moment of inertia with respect to longitudinal axis
$I_y$  mass moment of inertia with respect to transversal axis
$J_z$  added mass moment of inertia with respect to vertical axis
k1 Lamb's accession coefficient of inertia
k2 "              "            "                "
L  hull length at waterline
LCB  Longitudinal Center of Buoyancy
la  length of bow section of the body of revolution
In  effective ship length at waterline (body of revolution)
m  mass
$m_x$  added mass in the longitudinal direction
$m_y$  added mass in the transversal direction
$S_1$  slenderness = (cb*B/L)
T  mean draft at waterline
xbe  x-coordinate in which the sectional area = Abe
xbf  x-coordinate of center of bow fin
xcg  x-coordinate of center of gravity
ycg  y-coordinate of center of gravity
zcg  z-coordinate of center of gravity
xtf  x-coordinate of center of tail fin

xm  is the longitudinal distance from the hull fore-point
     to the origin of the body axes
CG centre of gravity

## Rudder

cr  rudder chord

brh  rudder height
hd  draft at rudder location
hr  effective draft of rudder
Ar  rudder area = cr*brh
Ra  rudder area = Ar
ARr  rudder aspect ratio = 2*hr2 /(Ar+Ari)
xr  x-coordinate of rudder center of gravity
$\delta$ rudder angle

## Propeller

ccp  propeller chord
dp  propeller diameter
Ap  lateral propeller area
ARp  propeller aspect ratio = dp/ccp
xp x-coordinate of propeller center of gravity
RPM Revolution Per Minute

## Skeg

cs  skeg chord
hs  skeg height
Asi  lateral hull area about skeg
Ass  skeg area
ARs  skeg aspect ratio = 2*hs$^2$ /(Asi+Ass)
xs  x-coordinate of skeg center of gravity

## Speed

V  forward speed of the CG
v'= v/V= -sin ($\beta$) sway speed (non-dimensional)
r'= r*LjV  rate of turn (non-dimensional)
u'= u/V= cos($\beta$)  surge speed (non-dimensional)
$\beta$ angle of attack

## Indices

$Q_H$'  non-dimensional bare-hull index on approach to a
    steady straight course
$Q_{HR}$'  non-dimensional hull-rudder index on approach to a
    steady straight course
$Q_R$'  non-dimensional rudder index on exit from a steady
    straight course
W  classification rules

## Coefficients
Y'  non-dimensional lateral force
Yv, Yr,
Yvrr,
Yrvv,
Yvvv,
Yrrr lateral force coefficients

Y$\delta$,

Y$\delta\delta\delta$ rudder coefficients of the lateral force

N' non-dimensional rotational moment

Nv, Nr,

Nvrr,

Nrvv,

Nvvv,

Nrrr rotational moment coefficients

N$\delta$,

N$\delta\delta\delta$ rudder coefficients of the rotational moment

**Others**

| | absolute value

p statistical average

$\Omega$ mass density of water

$\sqrt{}$ square root

$\sigma$ standard deviation

t time

th thickness

Fn Froude number = V/[Lg

$\theta$ hull afterbody angle

g gravity acceleration

## APPENDIX 11

## MAIN GEOMETRIC PARAMETERS OF THE SHIPS IN THE DATABASE

| TYPE | L (m) | B (m) | T (m) | Cb | Ra (m²) |
|------|-------|-------|-------|-----|---------|
| c1 | 115 | 16.3 | 3.02 | .739 | 10.419 |
| c2 | 148 | 19.399 | 8.4 | .68 | 39.782 |
| c3 | 137 | 18.5 | 8.07 | .699 | 34.273 |
| c4 | 156 | 19.5 | 7.179 | .619 | 36.962 |
| c5 | 134 | 18.399 | 3.959 | .66 | 15.919 |
| c6 | 162 | 21.399 | 9.779 | .699 | 53.868 |
| c7 | 112 | 15.8 | 7.32 | .729 | 25.415 |
| c8 | 152 | 20.6 | 4.019 | .68 | 20.775 |
| c9 | 114 | 16.399 | 3.27 | .67 | 11.183 |
| c10 | 115 | 16.3 | 4.25 | .739 | 16.62 |
| c11 | 140 | 18.199 | 7.82 | .68 | 33.938 |
| c12 | 41 | 8.199 | 1.419 | .56 | 2.037 |
| c13 | 132 | 20.6 | 4.019 | .595 | 14.618 |
| c14 | 114 | 16.399 | 3.27 | .667 | 12.343 |
| c15 | 145 | 19.5 | 8.019 | .676 | 18.001 |
| c16 | 132 | 18.199 | 3.549 | .696 | 14.286 |
| c17 | 115 | 16.199 | 3.02 | .745 | 14.004 |
| c18 | 137 | 18.5 | 3.609 | .688 | 14.01 |
| c19 | 144 | 19.3 | 8.259 | .664 | 18.076 |
| c20 | 157 | 19.6 | 8.25 | .616 | 18.609 |
| c21 | 133 | 18.6 | 8.099 | .736 | 16.523 |
| c22 | 148 | 19.399 | 8.4 | .381 | 19.702 |
| c23 | 138 | 18.8 | 3.959 | .66 | 19.31 |
| c24 | 123 | 16.699 | 6.65 | .765 | 12.942 |
| c25 | 129 | 18.199 | 8.429 | .666 | 16.756 |
| c26 | 115 | 16.3 | 4.95 | .677 | 14.09 |
| c27 | 205.74 | 28.956 | 9.753 | .589 | 36.221 |
| c28 | 278.892 | 32.308 | 10.363 | .76 | 52.168 |
| c29 | 123 | 16.5 | 7.28 | .685 | 14.057 |
| c30 | 137 | 18.5 | 8.07 | .696 | 16.7 |
| c31 | 140 | 19 | 8.349 | .707 | 19.515 |
| c32 | 106 | 15.599 | 3.04 | .638 | 11.346 |
| c33 | 148 | 19.399 | 7.94 | .65 | 19.977 |
| c34 | 115 | 16.3 | 4.95 | .68 | 22.769 |
| c35 | 86 | 12.5 | 5.66 | .729 | 7.788 |
| c36 | 94 | 13.699 | 4.15 | .689 | 11.312 |
| c37 | 161 | 20.399 | 9.019 | .689 | 31.948 |
| c38 | 122 | 15.9 | 7.62 | .76 | 15.803 |
| c39 | 122 | 15.9 | 8.199 | .76 | 18.007 |
| c40 | 137 | 18.5 | 3.609 | .689 | 8.902 |
| c41 | 144 | 19.3 | 8.259 | .66 | 19.031 |
| c42 | 157 | 19.6 | 8.25 | .619 | 19.428 |
| c43 | 133 | 18.6 | 8.099 | .739 | 32.965 |
| c44 | 138 | 18.8 | 3.959 | .66 | 19.126 |

| | | | | |
|---|---|---|---|---|
| c45 | 123 | 16.699 | 6.65 | .77 | 24.538 |
| c46 | 129 | 18.199 | 8.429 | .67 | 32.624 |
| c47 | 140 | 19 | 8.349 | .709 | 19.873 |
| c48 | 106 | 15.599 | 8.099 | .64 | 29.192 |
| c49 | 128 | 17.8 | 3.939 | .91 | 15.129 |
| c50 | 122 | 15.9 | 7.62 | .76 | 27.889 |
| c51 | 122 | 15.9 | 8.199 | .76 | 30.012 |
| c52 | 75 | 11.9 | 4.79 | .739 | 6.825 |
| c53 | 204.216 | 25.908 | 9.601 | .63 | 35.39 |
| c54 | 194.462 | 30.48 | 9.997 | .53 | 35.091 |
| c55 | 153.619 | 23.987 | 9.296 | .66 | 25.777 |
| c56 | 278.892 | 32.308 | 10.972 | .76 | 55.237 |
| c57 | 86 | 12.5 | 5.66 | .729 | 7.581 |
| c58 | 128 | 17.8 | 3.939 | .913 | 15.809 |
| c59 | 156 | 19.5 | 7.179 | .615 | 18.007 |
| c60 | 94 | 13.699 | 4.15 | .692 | 11.406 |
| c61 | 112 | 15.8 | 7.32 | .73 | 11.418 |
| c62 | 220.675 | 30.48 | 10.363 | .579 | 41.278 |
| c63 | 278.892 | 32.308 | 8.991 | .76 | 45.263 |
| c64 | 161 | 20.399 | 9.019 | .694 | 21.642 |
| c65 | 145 | 19 | 7.54 | .679 | 17.981 |
| c66 | 134 | 18.399 | 3.959 | .662 | 18.235 |
| c67 | 122 | 15.9 | 7.62 | .767 | 16.195 |
| c68 | 122 | 15.9 | 8.199 | .759 | 9.203 |
| c69 | 112 | 16.199 | 7.419 | .725 | 12.944 |
| c70 | 140 | 18.199 | 7.82 | .676 | 14.558 |
| t1-lng | 276.148 | 41.148 | 10.972 | .739 | 49.088 |
| t2-lng | 273.405 | 43.586 | 10.972 | .709 | 44.37 |
| t3-lpg | 233.172 | 39.898 | 12.74 | .76 | 47.591 |
| t4-lpg | 217.017 | 36.576 | 10.972 | .8 | 38.576 |
| t5 | 181.356 | 25.603 | 10.546 | .76 | 29.606 |
| t6-o | 216 | 30.6 | 10.32 | .811 | 31.351 |
| t7 | 201 | 28.199 | 10.83 | .805 | 28.718 |
| t8 | 193 | 26.5 | 10.31 | .801 | 27.483 |
| t9 | 192 | 26.5 | 10.32 | .796 | 27.405 |
| t10-o | 216 | 30.6 | 10.439 | .811 | 32.493 |
| t11 | 181.356 | 25.603 | 5.273 | .729 | 24.098 |
| t12 | 173.736 | 32.004 | 11.277 | .76 | 39.186 |
| t13 | 198.729 | 32.156 | 11.277 | .829 | 32.766 |
| t14 | 193 | 26.5 | 10.31 | .801 | 27.183 |
| t15 | 193 | 26.5 | 10.32 | .793 | 27.472 |
| t16 | 185 | 25.199 | 10.3 | .763 | 25.271 |
| t17 | 181 | 25.399 | 10.14 | .783 | 25.42 |
| t18 | 167 | 22 | 9.30 | .774 | 23.784 |
| t19 | 201 | 28.199 | 10.83 | .819 | 29.738 |
| t20 | 201 | 28.199 | 10.81 | .805 | 28.072 |
| t21 | 160 | 20 | 9.38 | .51 | 13.464 |
| t22 | 192 | 26.5 | 10.429 | .789 | 27.774 |
| t23 | 106 | 16.199 | 5.269 | .745 | 10.846 |
| t24 | 154 | 20 | 9.019 | .723 | 19.346 |
| t25 | 167 | 22 | 5.429 | .736 | 23.492 |
| t26 | 192 | 26.8 | 10.4 | .78 | 26.911 |
| t27 | 162 | 21.399 | 9.779 | .772 | 21.323 |

| | | | | |
|---|---|---|---|---|
| t28 | 198.729 | 32.156 | 11.277 | .8 | 34.693 |
| t29 | 227.076 | 33.223 | 12.192 | .79 | 41.859 |
| t30 | 232.562 | 35.356 | 10.668 | .79 | 38.405 |
| t31 | 260.604 | 32.247 | 12.192 | .819 | 57.191 |
| t32 | 260.604 | 32.247 | 12.801 | .819 | 60.05 |
| t33 | 260.604 | 32.247 | 11.277 | .819 | 52.901 |
| t34 | 232.562 | 38.1 | 12.161 | .8 | 41.398 |
| t35 | 232.562 | 38.1 | 9.753 | .79 | 33.201 |
| t36 | 235.915 | 39.319 | 12.192 | .8 | 43.489 |
| t37 | 235.915 | 39.319 | 13.106 | .8 | 55.655 |
| t38 | 235.915 | 39.319 | 10.058 | .79 | 37.16 |
| t39 | 289.864 | 47.365 | 15.849 | .819 | 78.561 |
| t40 | 289.864 | 47.365 | 17.373 | .729 | 77.957 |
| t41 | 185 | 25.199 | 10.3 | .76 | 26.676 |
| t42 | 106 | 16.199 | 5.269 | .75 | 10.613 |
| t43 | 154 | 20 | 9.019 | .719 | 19.447 |
| t44 | 192 | 26.8 | 4.139 | .589 | 11.128 |
| t45 | 192 | 26.5 | 10.4 | .79 | 27.955 |
| t46 | 289.864 | 47.365 | 7.62 | .739 | 32.667 |
| t47 | 263.347 | 52.73 | 13.106 | .8 | 53.429 |
| t48 | 263.347 | 52.73 | 10.058 | .75 | 41.481 |
| t59 | 288.036 | 50.596 | 19.69 | .8 | 87.794 |
| t50 | 192 | 26.8 | 10.4 | .589 | 27.955 |
| t51 | 216 | 30.6 | 10.3 | .81 | 31.147 |
| t52 | 288.036 | 50.596 | 17.16 | .79 | 75.624 |
| t53 | 319.308 | 50.017 | 19.263 | .819 | 88.881 |
| t54 | 319.308 | 50.017 | 9.265 | .77 | 52.191 |
| t55 | 330.708 | 51.816 | 19.964 | .819 | 96.526 |
| t56 | 192 | 26.5 | 10.599 | .8 | 28.492 |
| t57 | 167 | 22 | 5.429 | .739 | 12.695 |
| t58 | 162 | 21.399 | 9.779 | .699 | 22.181 |
| t59 | 201 | 28.199 | 4.889 | .739 | 12.777 |
| t60 | 167 | 22 | 12.199 | .739 | 28.523 |
| t61 | 162 | 21.399 | 5.94 | .699 | 13.471 |
| t62 | 216 | 30.6 | 5.269 | .76 | 15.936 |
| t63 | 330.708 | 51.816 | 9.601 | .77 | 51.438 |
| t64 | 324.916 | 53.004 | 10.607 | .77 | 53.35 |
| t65 | 350.672 | 54.955 | 21.183 | .819 | 114.993 |
| t66 | 350.672 | 54.955 | 10.18 | .77 | 54.62 |
| f1 | 111 | 16.399 | 4.78 | .567 | 17.627 |
| f2 | 113 | 15.9 | 4.4 | .561 | 11.698 |
| f3 | 111 | 17.399 | 6.799 | .569 | 18.87 |
| f4 | 113 | 15.9 | 6.799 | .56 | 19.21 |
| f5 | 119.481 | 23.164 | 4.389 | .46 | 17.305 |
| f6 | 155.996 | 24.597 | 4.998 | .589 | 27.292 |
| f7-cruis | 181.203 | 28.407 | 6.797 | .569 | 43.107 |
| bar1-tug | 91.44 | 18.897 | 4.724 | .989 | 15.551 |
| bc1 | 174.955 | 25.999 | 11.369 | .79 | 29.836 |
| bc2 | 269.748 | 42.976 | 14.325 | .78 | 52.168 |
| bc3 | 278.892 | 44.196 | 12.192 | .76 | 49.711 |
| bc4 | 278.892 | 44.196 | 15.849 | .839 | 79.831 |
| bc5 | 330.708 | 54.254 | 12.192 | .76 | 72.817 |
| bc6 | 330.708 | 54.254 | 16.154 | .829 | 96.483 |

# ABOUT THE AUTHOR

Professor Carmine G Biancardi, PhD, CEng, EurIng, (biancardi@uniparthenope.it), since 1982, he has been teaching Manouevrability of Ships, Safety of Ships, Naval Architecture and Total Ship Design at the Università degli Studi di Napoli "Parthenope", Italy. He has had several other appointments in International and European Research and Professional Bodies and Boards. He is a Member of SNAME since 1985 and since 1990 a Member of the Royal Institution of Naval Architects, UK and a Licensed Chartered Engineer. His past academic credentials include, among others, also the US Merchant Marine Academy, Kings Point, the University of Glasgow, UK, and the Australian Maritime College, Launceston, Australia, The Stevens Institute at Hoboken, NJ, Manhattan College, NYC, NY, The University of Tasmania, Australia. He has published more than 100 papers, books and chapters in the area of Engineering, Naval Architecture and Safety. In the last 30 years he has been researching, leading and coordinating over 50 international main research and innovative engineering projects.

His education includes: PhD in Ship and Marine Technology (Naval Architecture)at the University of Strathclyde, Glasgow, UK, 1993; Diploma of Communication at the Harvard University, USA, 1985; Msc (Dottore) in Nautical Engineering and Sciences, Universita degli studi di Napoli "Parthenope", Italy, 1983; Diploma of Master Mariner at the Istituto Tecnico Nautico, Italy, 1977

www.ingramcontent.com/pod-product-compliance
Lightning Source LLC
Chambersburg PA
CBHW081456170526
45166CB00008B/2444